Earthshock

Earthshock

Basil Booth and Frank Fitch

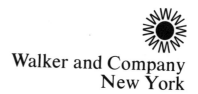

Walker and Company
New York

First published in the United States of America
in 1979 by the Walker Publishing Company, Inc.

ISBN: 0-8027-0593-6

Library of Congress Catalog Card Number: 79-64895

Printed in the United States of America

10 9 8 7 6 5 4 3 2 1

Contents

Acknowledgments

A large number of UN,UNESCO, multi-national, national, regional and private organizations are concerned with various aspects of natural disaster hazard and with aid and advice to those involved in natural catastrophes. In suggesting that these individual operations should be combined in one comprehensive United Nations World natural disaster agency we do not wish to minimize or denigrate in any way the immensely valuable work already being done in so many fields. The Scientific Event Alert Network maintained by the Smithsonian Institution of Washington, USA is now the major source of initial scientific information on new earthquakes, volcanic activity and meteorites. This and other information made available freely by the existing agencies is invaluable in disaster research. In preparing and writing this book the authors consulted and extracted data from hundreds of reference works, magazines, scientific journals, reports, research theses and articles, in addition to drawing extensively on their own unpublished research material. It soon became apparent that inclusion of any form of fully documented reference to the many sources of information used would badly disrupt the text and be an impossible burden to the general reader. Thus, with some reluctance, scientific references and footnotes have been omitted throughout. Nevertheless, the authors wish to acknowledge here their great indebtedness to countless other Earth scientists and writers, especially to their own research colleagues.

In preparing the illustrations invaluable assistance was obtained from Geoffrey Davenport, Malcolm Hobbs, George Reeve and other members of the Cartographic and Photographic Units at Birkbeck College. Plates 1, 24 and 25 were provided by NASA; plate 4 by the Press Association; plates 5 and 7 by the Swiss Tourist Office; plates 8, 9, 10, 11, 17, 18 and 19 by the Radio Times Hulton Picture Library; plate 12 by Solarfilma; plates 13, 14, 15 and 16 are from Lacroix, published by Masson, Éditeur, Paris; plates 20 and 21 by Popperfoto. The remaining illustrations came from the authors' own collections.

Finally, without endless patient assistance and encouragement from Stella Fitch this book would not have been completed. It also owes much to Susan Forster, whose growing family prevented her from being a co-author. Others whom the authors wish to thank particularly include Thelma Booth, Caroline Hooker and Melanie Gambold.

Preface

Every day, somewhere on Earth, people die as the result of natural disaster. We all hear or read – perhaps with momentary shock and fleeting compassion – TV, radio and newspaper accounts of the greatest, most dramatic or bizarre of these disasters, but the great majority go unreported other than locally. Today, most of us are quite unmoved by these repeated accounts of death and destruction, for in the second half of the twentieth century we have become over familiarized with meaningless bloody death as a result of the daily surfeit of real and fictional horror to which we are endlessly subjected. Nevertheless, the right to live is the ultimate human freedom. An otherwise caring society that, for lack of strong purposeful leadership unneccessarily exposes innocent people to the risk of death, whether it be from insufficiently restrained terrorism or other criminal activities or from the preventable consequences of natural disaster, is as much a failure as the most totalitarian of regimes. Natural disasters killed three-quarters of a million people in the year 1976; the annual average never falls below 20,000. Future natural catastrophes will undoubtedly kill many millions. While there is nothing man can do to halt the inevitable progression of natural events, there is a great deal that can and should be done to minimize their disastrous effects on human life and limb. Of course, like some criminal activity, natural disaster can be small, local in effect and best dealt with by personal and/or internal national effort. Local response to the threat of natural catastrophe is, however, totally inadequate in most instances. Natural disasters are not confined by petty political boundaries and the expensive expert knowledge required to give adequate warning of impending disaster and to assess accurately the dangers from natural events in progress is often far beyond the capabilities of those small poor states where life is most at risk. Perhaps in no field of human endeavour could effective international co-operation make a more useful impact and save more lives than by the successful prediction, risk assessment, control and subsequent mitigation of the worst effects of natural catastrophe. Clearly, it is not sufficient that the forces of world compassion be mobilized only after the event and with considerable delay. The fact that some nations – particularly Russia and China – do not allow free access to disaster areas and continue to restrict the availability of information for many years subsequent to the event severely curtails present international efforts to deal with the problem. It is our considered opinion that the United Nations should have a single, comprehensive world natural disaster control agency with certain overriding powers, able to investigate and prepare for all types of natural disaster anywhere in the world, ready to go in with the right kind of advice and aid immediately it is required, without political let or hindrance. To say that mankind's present fragmental, incomplete, piecemeal and politically hamstrung approach to natural catastrophe is foolhardy in the extreme is certainly no exaggeration.

In this book, we look at possible future natural disasters in the light of the knowledge gained over two-and-a-half centuries of Earth science research. James Hutton enabled geologists to explore the past history of our planet contained within its rocks by his demonstration that, in rock formation, 'the present is the key to the past'. The reverse of this dictum is also true: the past is equally the key to the future. Geologists have found that the history of the Earth is a record of innumerable natural catastrophes, some of gigantic proportions, any one of which may occur again and again. The record is such that we must ask ourselves the question: how long can man survive on Earth?

The Earth is a fascinating planet and proves to be ever more so the more we learn about it. We evolved here, the Earth is our home, but because both our immediate ancestors and ourselves have been around for less than a thousandth of the Earth's time-span, we have a very limited direct experience of the past history of our planet. We have some knowledge of the dangers that can arise from storm, fire and flood and from certain kinds of violent earthquake and volcanic eruption from the records of historical events, but we have no experience of many other, potentially greater hazards which geological exploration of the Earth and space exploration of the Moon and nearby planets reveals to be present. Undoubtedly the Earth has many unpleasant shocks in store for man. To assess this potential risk factor we must appreciate how the history of the Solar System and the Earth over the past 5000 million years can be examined and understood, and how the scant evidence of past disaster contained in the rock record can be read and interpreted in terms of cataclysmic prehistoric events that might well occur again in the future. Even a cursory study of the Earth reveals it to have been an exceptionally violent planet over its entire history. Are we seriously examining present and future natural hazards with sufficient vigour? There must be an international natural hazard research and control body to co-ordinate world effort in this field.

The modern view of the world is very different from that of our ancestors. In the Middle Ages the transitory nature of much of what we see around us was not appreciated and most natural disasters were attributed to supernatural intervention in the affairs of man. Geology, the science of the Earth and its history, had a very difficult, slow birth and it is only in the past hundred years that the full impact of the evidence that experienced geologists can extract from rocks has been fully appreciated. Direct knowledge of the Moon and our other intra-planetary neighbours has become available only in the last ten years. The ability to measure geological time and make accurate estimates of the ages of the Earth and other planets and of specific events in their past histories is vital to our evaluation of present and future natural dangers: these measurements depend upon our knowledge of radio-activity. Precise dating of rocks is, again, a very recent scientific development. We now know the Earth to be around 4600 million years old. If, in order to comprehend the vastness of this time span, we reduce it to a 24 hour time-scale, with each day representing 1000 million years, then it is seen that man evolved on earth around 14.28 hours on the fifth day and that the time is now only 14.40 hours on the same day! Geology reveals that every minute of these five days of Earth history has been filled with both gradual and, more often than not, violent activity. How do geologists read the 'record of the rocks'? Is it easy for the layman to see and appreciate some of this evidence for himself? What are the worst of the shocks the Earth may be holding in store for us?

Earthquakes appear to kill effortlessly: they are earthshocks in the direct physical sense. Millions have died in earthquake disasters in historical times. No year goes by without several great earthquake disasters being reported. The destruction of Lisbon in AD 1755 had a profound effect on European thought in the later part of the eight-eenth century. The experience of being near to the centre of a great earthquake, even outside a major city, is terrifying. The destruction caused by earthquakes in built-up areas can be devastating, as can the subsequent loss of life from pestilence and disease. Some of the most frightening accompaniments of some earthquakes are seismic sea-waves or tsunamis. The earthquake record of the Earth even over the past decade is one of unbelievable carnage. What causes earthquakes? Is there evidence in the record of the rocks that would suggest that even more violent earthquakes than those recorded in historic times could occur? Can earthquakes be predicted or prevented? What should we do to protect ourselves from this ever-present hazard if we live in a high earthquake risk area?

Mountains rise and crumble away, but the forces which cause them endure, and are always at work slowly but surely changing the very face of our planet. The rocky crust of the Earth is only a few miles thick and, like a cracked eggshell, is now broken and divided into a number of independent movable plates. Driven by deep, slow currents of hot rock the raft-like continental plates upon which we live are literally being taken for a ride. Throughout geological time the continental and oceanic plates have been dancing to and fro, at times marrying to produce huge supercontinents and then divorcing to produce a generation of smaller plates to carry on the dance. New plates are formed and old plates consumed as part of this apparently everlasting ritual. The recent discovery of plate tectonics and its mechanisms tells us why similar faunas are found on different continents separated by vast expanses of ocean, how volcanoes and earthquakes work and where continents once fitted together in different patterns in ancient times. The global pattern of earthquakes and volcanic activity is directly related to major plate boundaries, whether it is a constructive mid-ocean ridge or a destructive continental margin or island arc system. A study of plate movement tells us such things as how the Indian subcontinent broke away from the Antarctic and travelled north to ram Asia and throw up the Himalayas and Tibetan plateau and why, for example, evidence for past ice ages is now found in tropical zones. Clearly, it is the dance of our continents which is ultimately responsible for much of what happens on our planet. It governs the exact location of fossil fuels and ore deposits and, in general, the distribution of all natural resources, amenities and farmable zones upon which our lives depend. While the slow continental movements are not so dramatic as other events, their long-term effect can be devastating, producing volcanoes and earthquakes where today there are none, lowering large areas of land below sea level or raising the sea bed to form new land and, yet again, carrying other continents on an inexorable journey into the frozen wastes of the polar ice.

Volcanoes from time immemorial have been looked upon with superstition and awe. In ancient times they were regarded as gods or at least the abode of gods, where Vulcan forged thunderbolts for Jove or fashioned the weapons of Cupid. In 2500 BC a Greek island exploded in a tremendous volcanic eruption: at the same time the flourishing Minoan civilization disappeared and thus gave rise to the legend of Atlantis. In the year AD 1669 Mount Etna Volcano on Sicily erupted and is said to have been responsible for 100,000 deaths. Were either of these events to reoccur today, the death toll would be tremendous. In Sicily alone the death toll around Mount Etna could reach one or two million while giant sea-waves in the Aegean such as those which occurred in 2500 BC could devastate the now highly populated shores of Greece and Turkey to produce incalculable death and destruction. Nevertheless, although individual volcanoes can dramatically reduce populations in certain areas their immediate activity is unlikely to produce dramatic world wide effects on our planet. The long-term hazard of excessive and continued volcanism is rather pollution of the atmosphere to the extent that weather patterns are changed. Thick clouds of atmospheric dust resulting from explosive volcanism could prolong our winters and make them all the more severe, perhaps even precipitating another ice age. Recent research has demonstrated that cataclysmic eruptions from many of the world's so-called dormant or even extinct volcanoes could devastate areas half the size of the state of California, almost instantaneously wipe out cities like Portland, USA, Tokyo, Japan or even Rome, Italy. The governments of so many endangered countries are simply not aware of the dreadful possibilities, or if they are, they adopt the attitude that 'it can't happen to us'. But it can.

Throughout geological time continents have been repeatedly subjected to intermittent ice ages and wholesale flooding.It is a geological certainty that these processes will continue and even now there is a strong possibility of the return of another ice age in the northern hemisphere, devastating such cities as New York, Washington, London, Paris and Moscow. The conversion of good agricultural lands to frozen tundra wastes or their burial by thick ice and snow would produce famine on an unprecedented scale and force upon us a dramatic reduction of world population. Melting of an ice cap, on the other hand, would be equally disastrous, for it might raise the world sea level by up to 330 feet, flooding nearly all the major cities of the world and most certainly inundating over 75 per cent of our valuable agricultural land! Have the governments of the world sufficient political maturity to cope with earthshock of this nature? A recent suggestion that areas of polar ice should be broken up and melted by controlled thermonuclear devices or otherwise to produce supplies of fresh water is fraught with danger, for there may well be a point beyond which ice caps will fail to regenerate themselves. Whether by frozen death or flooding, the final effect on our presently divided world could be the same, the total annihilation of civilization as we know it, an unprecedented reduction in world population resulting from famine and pestilence and, perhaps, a more widespread return to a medieval way of life.

Every day of our life we are in danger of instant death from high speed missiles from outer space. Very small objects rain on the earth in their millions. Larger bodies collide with our planet at more infrequent intervals. Recent research reveals that the Earth may be struck by an extraterrestrial missile several miles across at least once every 50 million years and that the fall of smaller objects, called meteorites, is progressively more frequent as the size of the object decreases. Comets have always been regarded as bad omens: the devastating explosion that occurred on the Tunguska River in Siberia in 1908 has been attributed to collision with a minute member of this family – only a millionth of the mass of the average comet. Craters on Earth, like Meteor Crater in Arizona, resulting from the impact of medium sized meteorites are becoming reasonably well known. The mass spectrum of the debris available for space bombardment – mainly derived from the asteroid belt – can be estimated from our recently acquired knowledge of the pock-marked surfaces of the Moon, Mars, Mercury and Venus. The record of past impact events on Earth is rapidly destroyed by weathering and erosion and hidden or obliterated by sedimentation and crustal movements, but evidence is slowly accumulating which shows that our planet in its lifetime has been struck by as many rocky missiles as its neighbours. The number of structures confidently identified as 'astroblemes' or 'star wounds' resulting from ancient impact and cratering events increases annually. What effect would collision with a large asteroid, say, similar in size to that which caused Tycho Crater on the Moon 100 million years ago, have on the populated areas of North America? Could such an impact destroy life on Earth as we know it? Are we taking this remote but, nevertheless, very real hazard seriously? Is there a possibility of nuclear retaliatory strikes being ordered without justification if a major city in Russia or the west were to be unexpectedly destroyed by an extraterrestrial missile? Should an international space watch be mounted to warn of the approach of large intraplanetary bodies? If one were to be seen heading for the Earth is there anything man could do to avoid the disaster?

The answer to the question 'Can man and his planet survive?' must of necessity be based upon imponderables: it is, perhaps, partly dependent upon man himself and how he evolves socially over the next century. In view of the facts, it seems that man must settle his differences, come to terms with the forces which surround him and genuinely collaborate on a global scale to predict where and when disaster will next strike. In addition, mankind as a whole must at the same time be ready to alleviate suffering in a region of major catastrophe by directing resources where they will do most good.

List of Plates

List of Line Drawings

1 The Violent Planet

O God! that one might read the book of fate,
And see the revolution of the times
Make mountains level, and the continent, –
Weary of solid firmness, – melt itself
Into the sea!
William Shakespeare, *King Henry IV*

It is a popular delusion that the scientific
enquirer is under an obligation not to go
beyond generalization of observed facts . . .
but anyone who is practically acquainted with
scientific work is aware that those who refuse
to go beyond the facts, rarely get as far.
T. H. Huxley

The night of 23 March 1987 was still over much of the North Atlantic Ocean. Beneath a clear sky studded with brilliant stars the fast bulk container ship *Eldritch III* cut through the smooth oily waters at 30 knots, heading for its home port of New York. The single deck officer on watch, Larry Petersen, 30 years old and Brooklyn born, stared out from the warmth of the bridge. The unusual calm of the deep ocean seemed uncanny, although he had known such periods before. 'Calm before the storm' he thought, automatically running his eye over the large console in front of him to reassure himself that no warning lights were on. There were none: the great ship was in perfect condition, running smoothly under the guidance of its computerized control system. The chart display revealed that they were 350 miles ENE of Bermuda, some 800 miles out from the Narrows at the entrance to the Hudson River. There were no other vessels within a radius of 100 miles. Chief Engineer Ian McPherson, a Scot, and at 57 the oldest member of the crew, entered the bridge with a flask of hot coffee and two mugs.

'Dead quiet,' he remarked as he passed a steaming mug to Larry. One hour later the two men still sat in companionable silence watching as a pink dawn slowly crept up in the eastern sky. Suddenly the ship and the sea around it were flooded by an unnatural greenish-white light from the south that rapidly increased in intensity.

'What the hell is causing that?' Chief McPherson yelled as he ran across to the portside flying bridge to look out to the south. After a quick check of his instruments and the console, Larry joined him. High up in the southern sky a giant fireball, more than twice the size and brilliance of a full moon, was travelling rapidly towards the NNW.

'Fireball,' said the Chief, who had seen everything in his 34 years at sea. 'But larger than any I have ever seen or heard of.' They watched it crossing the sky for about ten seconds, during which time occasional enormous flashes and sparks of red and blue light illuminated what appeared to be a long dark trailing tail. As it passed over them the main fireball was seen to have a more cylindrical than round shape. By now glowing too brightly for them to stare at directly, it plunged over the horizon directly ahead of the *Eldritch III*. Darkness enveloped the ship for an instant then a gigantic flash of light occurred where the fireball had disappeared, so bright that they both instinctively turned their backs on it. For some seconds they could see nothing, then, as a weak pink sun rose through the misty eastern horizon, the long dark and serpent-like trail of the fireball became clearly visible across the sky.

'Fantastic,' said Larry, who had been rendered speechless by this display of natural pyrotechnics.

'Saw a fireball back in the summer of '76, off Harwich in eastern England,' commented the Chief, 'but this one must be the Daddy of them all.'

'What causes them?' asked Larry.

'Meteorites or comets tearing into the Earth's atmosphere,' replied McPherson, but Larry was not really listening; his eyes were riveted on events beyond the horizon ahead. With slow majesty a great mushrooming column of grey and white clouds could be seen rising right up into the stratosphere, a little way to the east of where the fireball had disappeared.

'Looks like a hydrogen bomb has gone off ahead, Ian,' he tried to say, but his words were lost in a great roar followed by a rumbling series of thunder-like detonations from the sky above as the sonic booms caused by the fireball and its trail hit them. The ship vibrated under this onslaught of sound but otherwise continued on its course without deviation. The Chief leapt across to the console to check the readings from his beloved engines. Larry quickly followed him, taking up his usual station again. He had just completed a full check of the ship when Captain Jim Barnes, a tall stoic New Englander, appeared on the bridge; like the Chief he had over 30 years' experience of the sea.

'What do you think, Ian?' the Captain asked. 'Has World War Three begun?'

'No,' replied McPherson hesitantly, then more firmly as he made up his mind, 'it could have been a nuclear spaceship crashing on Earth, but I think that it was a giant meteorite which passed over us and plunged into the ocean ahead.'

'How far ahead?' snapped Barnes.

'Few hundred miles,' McPherson replied. 'We saw the impact flash at 06.41 hours,' he added. Captain Barnes looked at him hard for a

second, then turned and pressed the emergency alarm button.
Moving to the Captain's chair he sat down and picked up his hand
microphone.

'Now hear me, this is your Captain speaking,' he said quietly into the
handset. 'I want all crew members to their emergency stations, all hatches
battened down and everything movable stowed away or lashed down.
And I want all this done within 10 minutes. Go!' Larry turned to the
Captain, his face expressing his puzzlement.

'A giant sea wave may hit us, Mr Petersen. Alternatively, the very least
we can expect is a patch of very bad weather,' explained the Captain.
'Look at those clouds.'

Larry turned his eyes to the horizon once more. For a good 30° of arc
the distant sky was filled with boiling, billowing cloud, black as ink below,
white as ice above, with an enormous anvil-shaped plume rapidly
extending far into the stratosphere. Even as he looked, a thin line of black
and grey clouds began to slowly spill out across the horizon at the foot of
the tall mushroom-like column. At 07.20 hours, just as Captain Barnes
picked up his microphone to relay another message to the crew, three
immense detonations shook the entire ship. A few minutes later, Chief
McPherson, who had been below, reappeared with his large pocket watch
in his hand.

'The impact area must be about 390 miles ahead of us,' he informed the
Captain. 'Expect any tsunamis within the next hour or so.' He looked
thoughtfully at Jim Barnes for a moment. 'The coast of New England is
due for a nasty time, if there are large tsunamis,' he added, knowing that
the Captain's wife and family lived in an old rambling frame house close to
the shore in Rhode Island. Jim Barnes thought about this for a while, then,
suddenly making up his mind, he picked up the telephone built into the
arm of his chair and told the radio operator to get him Head Office on the
radiotelephone. In a few moments he was speaking to the duty manager in
the shipping company's New York office. In response to his questions he
soon learned that the whole Eastern Seaboard of the United States, from
Boston to Norfolk, Virginia had been illuminated by a brilliant flash of
light just before dawn and that this had been followed by the noise of three
tremendous explosions around 7.25. Nobody knew the cause: there was
much speculation on the early morning news programmes, but the official
attitude in Washington was 'No Comment'. The weather in New York was
fine, but visibility was poor to the east because of a thick bank of sea mist
behind which heavy storm clouds could be seen building up. Captain
Barnes reported the observations and conclusions of his Chief Engineer
and himself and emphasized their belief that a destructive tsunami might
strike the coast. The voice from the 'phone was polite but uninterested as
it thanked him for his concern; Barnes then asked to be connected with
the office telephone exchange, and within two minutes he was speaking to
his home.

'Whatever you are doing, drop everything, get everyone in the car and
drive as far inland and onto high ground as quickly as you can,' he ordered
his wife after the briefest explanation and a reminder to her of the damage
done by a tsunami that they had both seen in Hawaii some years before.

The time was 7.40 a.m. Before she could argue he slammed the 'phone down and turned to Larry.

'Your parents in Brooklyn, Mr Petersen?'

'They live on the tenth floor of a new building, sir, should be OK,' he grinned.

'Hurumph,' replied the Captain, trying not to let the chilling fear he felt for the safety of his wife and children show on his face or in his manner. He turned to look at the horizon once more. He blinked and looked again: a frightening change had occurred over the last few minutes. All the western and northern horizon in front of him was now closed in by a vast bank of angry cloud with lightning playing intermittently within it. In front of the cloud, the sun still shone sufficiently brightly to illuminate a thin band of dancing silver lying directly along the horizon and over the following twenty minutes everyone watched the silver band grow into a huge wave over 200 feet high, racing towards them at close on 300 m.p.h. At 7.50 a.m. the bow of the ship began to rise.

'Full ahead, Mr McPherson,' the Captain said quietly. Tapping him gently on the shoulder he replaced Larry Petersen at the main console, where he sat down heavily, leaning forward gripping the seat arms fiercely. The water ahead was broken and torn by random gusts of wind, but otherwise rose relatively smoothly in front of the ship to a mountainous crest some 80–100 miles away, but approaching fast. By 8.10 a.m. the *Eldritch III* had struggled sluggishly to the crest of the enormous wave and was poised to plunge down into what appeared to be a cauldron of hell. Immediately ahead was a great trough, as deep as the mountainous wave on which the ship floated, and in which the sea had a uniform black bleakness. Behind this, further great waves covered with tumultuous foaming water could be seen approaching beneath the front of a great and incredibly high wall of black writhing cloud that raced forward to envelop them as they slid down into the trough. The sun was suddenly shut off by the fast-spreading clouds in the stratosphere and the ship raced forward into the stygian gloom, the engines now turning at only a quarter their normal speed.

For the next four hours, the *Eldritch III* was shaken, tossed, lashed with rain and hail and flooded by great seas in the worst typhoon that any of the crew could remember, again and again rising and falling beneath the passage of giant sea waves. Nevertheless, none of those that followed were anything like the first in size. The second largest was not more than 100 feet in height, the remainder not more than 30 feet, except where the wind piled crossing wave patterns one on top of another. The electrical interference was so bad for most of this time that radio communication was impossible, although the Radio Operator twice reported hearing faint and undecipherable distress calls. At midday, the storm ceased as suddenly as it had begun and the *Eldritch III*, now running normally once more, sailed into a relatively calm sea beneath broken cloud. Captain Barnes sighed and turned to relinquish the main control console to Larry Petersen. He picked up his hand microphone to order the end of the emergency. Before he could speak, however, the radio operator burst onto the bridge, white-faced, his eyes wild.

'The whole of the Atlantic seaboard of the States has been swept away,' he shouted, 'The tidal wave hit them between 8.30 and 9 – New York, Washington, Boston – they've all gone! The Vice President was in LA, he's taken over and declared martial law: it's a national disaster!'

Throughout the Eastern Seaboard of America the story, when it was pieced together, was the same. The coast between Nantucket Island and Cape Hatteras had been completely inundated and devastated by the three great waves. No buildings remained undamaged and there were very few survivors. The width of the devastated strip depended upon the local topography. Where the land behind the shore was exceptionally low-lying, the swash of the first giant wave had penetrated up to 50 miles inland; elsewhere, high ground had contained the surge to the actual coast. Away from the ocean shore destruction was confined to major river valleys – enormous surges had run long distances up the Hudson, Delaware, Susquehanna, Potomac, James and similar rivers, causing terrible damage and loss of life in cities such as Albany, Philadelphia, Harrisburg, Baltimore, Washington and Richmond. Nevertheless, the greatest damage was confined to the cities and towns that were situated actually on the ocean: Norfolk, Virginia; Atlantic City; New York and the coastal cities of Connecticut and Rhode Island. Cape Cod, and the islands of the Atlantic Coast – Nantucket, Martha's Vineyard, Long Island – and many low coastal areas in Delaware, Maryland, Virginia and North Carolina were all almost completely cleared of any trace of former human habitation. Boston suffered badly, but being protected to some extent by the Cape Cod peninsula was not so completely destroyed as were the lower-lying parts of New York.

One other place which suffered terribly was Bermuda, from which there were no survivors. Outside the limits of the worst devastation, Halifax and Charleston were badly hit, great damage occurred in Florida, the Bahamas, Newfoundland, the Azores and elsewhere, and many ships and smaller vessels were lost at sea. The *Eldritch III* picked up three survivors from an ocean racing yacht on the day after the impact and was nearing New York in a sea littered with flotsam and jetsam before it was diverted to the St Lawrence on orders from the owner, who had been on holiday in Switzerland. Captain Barnes and his crew learned with relief that the President was, in fact, alive and actively directing the rescue and rehabilitation programme: the early reports of his death arose from the initial failure of all communication systems in Washington-and along the Eastern Seaboard. The glad news that the Barnes family had escaped the destruction and disappearance of their home by travelling some 35 miles inland before the tsunami struck the Rhode Island coast was later overshadowed by confirmation from Brooklyn that no brick or stone was left of the new building in which Larry Petersen's mother and father had lived. In all, over 10 million lives were lost on the shores of the Atlantic Ocean in a few moments of terrifying horror. The world would never feel safe again.

*

This story is, of course, science fiction: such a horrifying natural disaster has not happened yet and, indeed, may never happen. Nevertheless, we know that large meteorites capable of causing devastation on this, or on an even greater scale, do strike the Earth at rare intervals. Only 70 years ago, in 1908, an extraterrestrial impact in a remote part of Siberia caused an explosion equivalent to the detonation of a 30-megaton atomic device, and, since then, great scars caused by ancient meteorite impacts have been identified in various places all over the world. Thus, unlike much of the science fiction we read today, our story of the destruction of New York and much of the East Coast of the USA by a tidal wave induced by meteorite impact tells of a scientifically *possible* future event.

The purpose of the story is to demonstrate the extent of the havoc that might be caused by just one of the many kinds of scientifically possible future disaster. In this book we shall examine all the major natural hazards of life on earth – the large-scale extraterrestrial bombardment, earth movements, volcanism, sea level, climatic and

Fig. 1   Roaring in at terrific speed from 'outer space' some 25,000 years ago a million ton meteorite strikes the Earth in Arizona. Incandescent on the outside and beginning to fragment already as a result of its passage through the Earth's atmosphere, on impact it was completely blown to pieces and partly vaporized in a great explosion which excavated a crater 600 feet deep and nearly 4000 feet across.

other changes – that might kill millions or so change the environment as to make life as we know it difficult or impossible to maintain.

Geologists know that innumerable terrible earthshocks, each far greater than any that have occurred in historical times, have struck our planet from its very beginnings and will continue to do so until it ceases to exist, for the geological history of the Earth is an incredible story of seemingly endless change and natural violence. For thousands of millions of years, the thin surface skin of rocky scum that forms the continents we live on has been repeatedly crumbled, torn apart and reconstructed by deep-seated forces. Subjected to the everlasting violence of volcanism, earthquakes, subsidence, upheaval, folding, faulting, storm, hurricane, drought, fire, flood, glaciation and bombardment from space, their substance continually augmented by additions of molten rock from below, the shattered fragments of ancient continents are slowly swept around our mobile globe to collide and rejoin each other to form new lands – which are themselves destined for dismemberment in later years as the inexorable cycle of change and violence continues. Geologists expect the Earth to remain in this 'active' phase of its life for thousands of millions of years to come.

The rocks of every continent contain a record of innumerable major and minor geological events, every one of which would be a catastrophic disaster to mankind if they occurred today – and it is scientifically certain that such events *will* occur again and again in the future. Mankind must be prepared for the occasional advent of natural disasters of such colossal magnitude that in each event millions will die and whole civilizations collapse. This is not an irresponsible prediction. While nobody can say precisely what the future holds in store for us, after nearly 250 years spent in a careful and painstaking examination of the history of the Earth, geologists are now in a position to offer definite advice on the inevitability of a variety of catastrophic *earthshocks* that must occur again *some time in the future* and, on the other hand, to dismiss some of the wilder flights of uninformed prediction. Withdrawal of North Sea oil, for instance, could not cause a gigantic inward tilt and subsidence of the surrounding land areas, nor would it be possible for man to 'knock the Earth from its orbit' as a by-product of atomic warfare. It is obvious that man's selfish and unthinking activities can do serious harm to the environment and make life very much less pleasant for generations to come, but the total destructive power that we can at present generate is still incredibly puny when compared with the awesome forces of the natural world. Nothing man can command can compare with the energy released during the great Assam earthquake of 1950, for example, or match the violence of the paroxysmal eruption of the Bezymianny in Kamchatka in 1956, yet such events are but small happenings in the continuous flow of the natural events on our planet and, when viewed in the context of the Earth's history, fade into insignificance in a multitude of equal and occasionally far greater events.

Table 1.1   Comparison between various types of natural disaster

Type of disaster	Greatest known loss of life	Estimated maximum loss of life in a future disaster
Flood due to rain	In June 1931, in the Honan Province, China, the Yangtze and Yellow River floods killed between 1 and 2 million people.	2–3 million
Earthquake	On 24 January 1556, in Shensi Province, China, an earthquake killed 830,000.	1–1.5 million
Tsunami of seismic origin	Bay of Bengal, 1876, 'tidal waves' killed 215,000.	0.25–0.5 million
Tsunami of volcanic origin	Krakatoa, 27 August 1883, 'tidal waves' killed 36,417.	100,000–200,000
Volcanic eruption	Etna erupted in 1669 destroying Catania, Sicily reputedly killing 100,000.	1–2 million
Typhoon or hurricane	On 8 October 1881 a typhoon at Haiphong in Vietnam killed 300,000.	0.5–1 million
Storm	On 26 November 1703 a storm in the English Channel killed 8000.	10,000–20,000
Landslide	On 16 December 1920 a landslide killed 200,000 in Kansin Province, China.	0.25–0.5 million
Avalanche	On 13 December 1941 an avalanche at Huaras, Peru killed 5000.	10,000–20,000
Pestilence, plague or other epidemic	'The Black Death', a prolonged outbreak of bubonic plague in the fourteenth century killed 25 million people in Europe.	many millions

Type of disaster	Greatest known loss of life	Estimated maximum loss of life in a future disaster
Famine (resulting from drought and other causes)	In Bengal in 1943 and 1944, 1.5 million died in a famine.	many millions
Asteroid impact	Not known in historical times.	hundreds of millions
Return of an ice age or a significant rise in world sea level	Not known in historical times.	hundreds of millions through famine

Other possible disasters not considered above include: uncontrollable growth in world population; exhaustion of energy supply; exhaustion of raw material sources; uncontrollable spread of a genetic defect, either naturally or as a result of 'genetic engineering'; uncontrollable swarming of insects or some other species to the exclusion of man; thermonuclear holocaust; beyond redemption pollution of the seas, the land or the air; runaway biological warfare; failure of man-made structures such as bridges, dams, tunnels; loss of control of man-made mechanical aids, robots, space stations, etc.; invasion from space (not necessarily by sapient beings – invasion by an unknown virus could be equally disastrous to man).

Our very limited direct experience of natural hazards is a major problem. Man has been on Earth for less than the last thousandth part of the Earth's life span and detailed scientific records of the disasters suffered by our ancestors are few. The oldest natural catastrophe known to have affected man was a flash flood which raced down an apparently dry Ethiopian river bed some 3 million years ago, drowning a small family group of early men, women and children caught in its path. The pitifully scattered remains of their skeletons, buried by old river and lakeshore sediments, were found in the Afar during the summer of 1975. Unexpected flash floods still claim many victims every year: in 1952, the Lynmouth disaster claimed 34 victims in England (Plate 4), and in 1972, 237 people were killed by a flash flood that tore through Rapid City in South Dakota, showing that we have yet to learn to avoid this danger, even after all these years' experience of its effects. Even in our technologically advanced society, small uncontrolled natural disasters kill about 20,000 people every year: numerous sudden deaths from snow avalanche, landslip, hurricane, storm or tsunami are reported in the press of the world, and volcanic disaster and earthquake together claim many thousands of lives annually. In particularly bad years the death toll may reach

Fig. 2 'In the midst of life we are in death.' When the volcano Vesuvius erupted in AD 79 the people and animals that were trapped in the previously delightful Roman town of Pompeii died horrible deaths. Most of the 20,000 inhabitants of Pompeii escaped in time, but some 2000 people were left behind: remarkably detailed casts of a number of these people and their animals have been found preserved in the enclosing ashes. It is probable that the rain of hot tephra became so heavy and incessant during the night of 24 August that it was impossible for those remaining in the town to leave the shelter of the houses. Cowering inside, they were either buried alive when the houses collapsed under the weight of the hot ash and pumice fragments continually falling on them, or were asphyxiated by the noxious fumes being given off by the tephra. The destruction of the popular towns of Herculaneum, Pompeii and Stabiae by the eruption of a volcano formerly thought to be extinct came as a great shock to the Roman populace.

between 1 and 2 million, and we have to accept that at any time even greater catastrophes that could each kill tens or hundreds of millions may overtake us.

A search through the historical records of the last 1000 years reveals that man has been subjected – even in this relatively short time interval – to an incredible sequence of horrifying natural disasters. If we disregard famine, pestilence, plague and other forms of biological disaster, the greatest single catastrophes known to have occurred during this period would appear to have been in China. The earthquakes of AD 1556 and 1976, in Shensi Province and around Tientsin, respectively, each killed between a half and one million persons, and the 1931 floods in Honan Province caused one-and-a-half million deaths. To those directly involved, however, every natural disaster, even if the loss of life is small, is equally terrible, as the real-life examples on the next pages will show.

One splendid day in the autumn of AD 1619, most of the men of Dughabad, in the province of Kuhistan, Iran, were out in the fields gathering what promised to be a fine grain harvest. In the town, the marriage of the daughter of one of the foremost citizens was being celebrated at a large and affluent wedding party. Suddenly, with virtually no advance warning, Dughabad was shaken by a severe earthquake. Within minutes the previously flourishing town was reduced to a heap of rubble and dust. Seventy women were killed at the wedding party alone, though the bride escaped by throwing herself under the massive cross-beam of the doorway to the house. In all, between 700 and 800 people, mostly women, lost their lives in Dughabad during the earthquake – the wedding feast became a funeral. When rescue operations were complete and all the dead had been buried, no building or wall was still standing. Although the town was rebuilt over a number of years, it never regained its former wealth and importance. The men and women of early seventeenth-century Dughabad had suffered a tragic and, to them, incomprehensible earthshock!

In AD 1700, after four years of difficult construction work, the English engineer Henry Winstanley (formerly clerk of works to King Charles II) completed his masterpiece – hailed as one of the wonders of the world – the first Eddystone lighthouse. It took the form of a massive wooden polygon 100 feet high on a stone base, erected to mark a group of dangerous rocks lying in the seaway between Start Point and the Lizard in the English Channel. Nine miles from the Cornish coast and 14 miles SSW of the important naval port of Plymouth, the Eddystone rocks are only visible at ebb tide and had been a serious danger to navigation in the Channel since Roman times. Winstanley and his contemporaries were to enjoy their triumph of man over nature for no more than three years, for a great earthshock was in store for them. Beginning at one o'clock on the morning of 27 November 1703, the Channel and southern England were struck by the most violent and sustained storm in history. Thunder, lightning, hail, rain and severe winds gusting to hurricane force ravaged the country until early evening. There were record high tides in the Thames and Severn estuaries, causing floods. At sea, whole fleets were cast away and many smaller ships were either wrecked or simply vanished. The Eddystone lighthouse was destroyed and its designer killed. All along the Channel coast was utter devastation. At London Bridge, the Thames was blocked by wrecked shipping. The cities of London and Bristol looked as though they had been subjected to destructive warfare. Houses, mansions, church steeples and thousands of trees were toppled. The lead was ripped from the roofs of buildings and blown away like scrolls of parchment. Tiles flew dangerously into streets, chimneys and battlements fell. The great oaks planted by Cardinal Wolsey at St James's in Henry VIII's time were uprooted and Queen Anne's new rows of limes, acacias and elms were blown down in their hundreds. In all some 8000 people died, from the Bishop of Bath and Wells and his wife, who were killed in their bed by the fall of a chimneystack, to an 'abundance of brave men irrecoverably lost at sea'. The whole British nation was shocked: no one in southern England would ever forget this great storm.

In Caracas, Venezuela, 26 March 1812 was a remarkably fine, hot day. The air was calm and the sky unclouded. It was Holy Thursday and most of the populace were assembled in the many churches that were then a prominent feature of the city. There was nothing to presage the advent of a terrible calamity. Without warning, at seven minutes to four in the afternoon, the first sign of the great earthquake to follow was felt. This foreshock was sufficiently powerful to make the bells of the churches toll; it lasted for five or six seconds, during which time the ground was in a continuous gentle undulating motion, seeming to heave up and fall back like the surface of a boiling liquid. When these ground movements ceased, it was at first thought that the danger had passed, but, before the fear generated by the foreshock had abated, suddenly, a tremendous subterranean noise was heard, resembling cannon-fire or the rolling of thunder, even louder and more continuous than the thunder heard during the fiercest of tropical storms. The people once more were panic stricken, but could do no more than pray. The dreadful noise preceded further, extremely violent, up-and-down motions of the ground lasting for about three or four seconds. These up-and-down motions had an apparent amplitude of several feet and were followed by longer-period undulations of decreasing intensity that appeared to be travelling both north–south and east–west. No building then existing could withstand this combination of ground movements and the city of Caracas was entirely overthrown. Between nine and ten thousand people were buried beneath the ruins of churches and houses; in all, some 90 per cent of the buildings in that city were completely destroyed. There were many aftershocks. In the country surrounding Caracas, great landslips scarred the mountains, lakes were partially drained and new springs gushed out at a number of formerly dry places. The famous Prussian naturalist, Baron von Humboldt, reported that 20,000 persons perished in Caracas and the adjacent province as a direct result of this unexpected and calamitous earthshock.

The tallest mountain on the North Island of New Zealand is Ruapehu (9170 feet), an active andesite volcano (see Chapter 5) that has snow on its summit for most of the year. The large summit crater of this composite ash-lava-lahar volcano contains a crater lake that can be as much as 2000 feet across. Periodically, the release of hot volcanic gases into this lake from below increases its temperature and acidity until it begins to steam gently all over and its surface becomes covered with a foul-smelling yellow sulphurous scum. After a summit eruption in 1948, the outlet from the lake became blocked with a mixture of volcanic debris and ice. During the succeeding years the level of the water in the lake gradually built up until, on the night of Christmas Eve 1953, assisted by a minor rise in volcanic activity in the summit region, the accumulated waters of the crater lake burst through the barrier of ice and debris, releasing about 600 million gallons of hot acid water almost instantaneously. The lake level fell by 30 feet. A terrifying volcanic *lahar* (a great churning mudflow of ash and rock from the collapsed crater wall, volcanic debris and boulders from the side of the mountain and mud from the bottom of the lake, all lubricated by the acid contaminated water) rushed into the head-waters of the Whangaehu River, which drains Ruapehu on its southern side. A few miles downstream, the main railway line from Auckland to Wellington crosses the river at the Tangiwai Bridge. On 24 December 1953, the northbound Christmas Eve express thundered through the night. The express and the lahar were approaching each other on convergent courses. The lahar reached the Tangiwai Bridge five minutes before the train, effortlessly sweeping away the lines in a cascade of turbulent mud, rocks and water. For a few moments life went on as before on the train – then, as it reached the Whangaehu River, it plunged straight into the thick, swirling, rock-full torrent of the lahar. The engine and carriages were swept away downstream and many coaches were smashed like boxwood. In all, 151 people died: for these particular victims, death must 'have come in an incomprehensibly terrifying manner on the eve of their anticipated joyful Christmas holiday.

The largest earthquake in over a century struck the 'big island' of Hawaii at 4.48 a.m. on 29 November 1975. It had a magnitude of 7.2 on the Richter scale (see Chapter 4) and was centred below Kalapana on the south-eastern coast of the island. A foreshock at 3.36 a.m. woke most residents, including 32 people who were camping near the beach at Halape, a few miles west of Kalapana. The foreshock triggered rock-falls from a hill behind the campsite and a few campers moved closer to the beach, but most went back to sleep. Further rock-falls occurred when the main earthshock followed just over an hour later. At the onset of this earthquake the campers were unable to stand because of the violent shaking of the ground. Within a minute or so they were caught by the rapidly rising sea, which quickly became a surging wave. A second, larger and more turbulent wave followed behind the first and washed trees, rocks and people into a 20 foot deep crack in the ground, within which the campers were churned uncontrollably by the foaming waters. A survivor described the experience as like 'being inside a washing machine'. One person was drowned or battered to death and another washed out to sea and lost. The maximum height reached by this seismic shock wave or tsunami was 48 feet; it swept inland for some 325 feet from the beach. The surviving campers were very lucky to be alive! Elsewhere on Hawaii there was considerable damage to roads and property. A small sympathetic eruption occurred in the summit caldera of the volcano Kilauea.

Natural disasters can be sudden and dramatic, as in the examples quoted above, or they can be more insidious, creeping up on us slowly, their full horror only becoming apparent as their cumulative effects begin to cause an accelerating number of deaths. Climatic change is one of the main offenders in this latter category. In the mid 1970s we were in the grip of just such an insidious earthshock in the tropical regions of Africa and Asia. Failure of the usual rains led to spreading desertification in many previously fertile areas, especially, for example, along the southern margin of the Sahara Desert in Africa. Slowly increasing malnutrition, starvation and, finally, death, were the lot of numerous primitive peoples and animal herds caught in the toils of this less dramatic form of earthshock. In densely populated regions, such as India and China, climatic accidents of quite short duration can lead to frightening famine if the main food crops are adversely affected. In this respect, floods caused by an excess of rainfall can be as damaging to agriculture as rainfall failure.

Other slow earthshocks include such locally disastrous events as the remorseless advance of glaciers and ice sheets, the steady erosion of shorelines by the ocean, the gradual subsidence of large areas of land beneath the sea and the slow changes caused by mountain building movements in regions of continental collision or along the junctions between oceanic and continental crustal plates. Examples illustrative of insidious earthshock in historical times are legion – the mid twentieth-century retreat of certain glaciers in central East

Greenland has revealed the remains of small medieval communities that were completely overwhelmed by an ice advance that began in the middle of the seventeenth century; the grandiose capital city of Fatehpur Sikri, 'the city of victory', built by the Muslim Emperor Akbar south-west of Agra in India between 1569 and 1572, had to be abandoned within 50 years of its completion because its water supply failed; the renowned English medieval town of Dunwich, formerly the capital city of the independent kingdom of East Anglia, had over 200 churches in the fourteenth century but was progressively undermined by the sea until only a few houses now remain. Walking today on the shingle beach at Dunwich on a windy winter day, the melancholy atmosphere – it is said that during storms the bells of the lost churches can be heard ringing – is heightened by the occasional find of washed bones of long-dead monks from the graveyard of the famous monastery.

In 1830, it was discovered that the platform of the Roman Temple of Jupiter Serapis, near Pozzuoli in the Bay of Naples, lay one foot beneath high water mark, yet three slender marble columns rising from this platform were pitted by the borings of marine shellfish to a height of 24 feet. From historical accounts, it was known that the Temple had stood on dry land during the reign of the Emperor Marcus Aurelius in the second century AD. Clearly, between the time of Marcus Aurelius and 1830 the Temple had subsided beneath the sea to a depth of, at least, 24 feet, and had then been re-elevated, all without a disturbance of sufficient violence to topple the marble columns. To the early nineteenth-century geologists who examined the Temple, this was compelling evidence that long-term slow changes in the relative elevation and distribution of land and sea do take place. Modern detailed surveying of much of the surface of the globe reveals that subsidence and upheaval of the land and continental drift relative to the axis of rotation of the Earth are continuously in progress at rates which vary from very slow to quite noticeable movements in certain areas. Sometimes, man's activities can make things decidedly worse. It has been suggested, for example, that an excessive use of underground water reserves may be accelerating the gradual subsidence of the beautiful Italian city of Venice beneath sea level.

Earthshock is not always to man's disadvantage: past volcanic eruptions in many parts of the world have produced new land of invaluable fertility. Elsewhere, whole islands and countries, like Hawaii and Iceland, are the product of repeated volcanism and earthquake. Nevertheless, direct involvement in violent natural events such as earthquakes, lahars or meteorite bombardment is inevitably catastrophic for the individuals concerned, and the long-term effects of insidious earthshock can often be equally disastrous for whole human populations.

How can we assess the future risk? The problem is that any realistic comprehensive assessment of the future risk to man from the various forms of earthshock must supplement our limited modern experience – and the poor, incomplete and often factually distorted record of disasters that have occurred over the past few thousand years – with evidence from the previous history of the Earth. For although our forbears go back many millions of years, our nearest ancestors – those we should recognize as men – have been on Earth only a very short time in terms of the history of the universe. Our genus, *Homo*, probably evolved in Africa some 4 to 5 million years ago; a very long time by an individual man's promised life span of three score years and ten, but a very short time compared with the age of the Earth itself – over 4600 million years! The record of modern man's direct experience of the vicissitudes of nature goes back, in writings, carvings and cave paintings, no more than 30,000 years. Thus our knowledge of direct human involvement in the great natural upheavals, processes and cataclysms of the Earth is very scanty and is, at its best, a poor and unrepresentative sample of the Earth's past activity. Obviously such a short time interval is an even less representative sample of the possible extraterrestrial disruptions to life on Earth.

In these circumstances, how long can man survive on Earth? Is our eventual extinction inevitable? Will we be destroyed by the forces of nature or could we learn to control them? Are we seriously examining present and future earthshock hazards with sufficient vigour? Should there be an international natural disaster research and control body to co-ordinate world effort in this field? To scientists who study the Earth these are more important questions than the petty political squabbles that fill the pages of our daily newspapers. Serious discussion of these matters and any useful evaluation of the risk factor require a scientifically unbiased understanding of the natural processes of the Earth and a full knowledge of the long history of repeated violence and insidious earthshock that is contained in the Earth's 'record of the rocks' over its entire time-span of 4600 million years. But what is certain is that our documented experience of the great terrestrial earthquakes and volcanic explosions that have occurred in historical time, such as the Tientsin earthquake of July 1976, that killed over half a million people, and the complete volcanic disruption of the island of Krakatoa in 1883, gives us no more than a foretaste of the scale of natural violence that is possible on Earth.

2   The Record of the Rocks

Tomorrow, and tomorrow, and tomorrow,
Creeps in this petty pace from day to day,
To the last syllable of recorded time;
And all our yesterdays have lighted fools
The way to dusty death.
              William Shakespeare, *Macbeth*

A major problem in comprehending Earth history, experienced alike by both scientists and laymen, is the nature of the time scale involved. As individuals, we are familiar with events that follow each other in seconds, minutes, hours, days, weeks, months, years and even centuries, but the jump from this human time scale to that of Earth processes, in which the events follow each other at intervals of millions or hundreds of millions of years, is not easy. When we take a favourite walk in the countryside, it is not difficult to imagine that the hills, mountains and streams we see have always been there, but this is not so. A knowledge of the Earth's history as derived by geology tells us that all we see is transitory. Even the major elements of the world's present-day geography are no more than the ephemeral impression of one moment of time. Recent studies, for example, have shown that the rock crust beneath the floor of the great oceans is nowhere older than 250 million years and is mostly much younger. In many parts of the world new oceanic crust is forming now! Again, we now know that the apparently 'everlasting' granitic rock crust that forms the upstanding continents – and is arbitrarily divided into our jealously guarded national states – has, in fact, split apart and reunited in a never-ending dance over the past 2500 million years.

If an observer in outer space had been able to make a time-lapse film of the Earth over the 4600 million years of its life, in its speeded-up projection we would acquire a very different view of the Earth from that obtained on our tranquil walk in the countryside; in such a time-lapse film the Earth would appear a very violent mobile globe – completely unlike the virtually dead planets Moon, Mercury or Mars whose surfaces are now familiar to us from the photographs taken by astronauts and space-probes. Beneath the ever-changing swirl of storms, hurricanes and cyclonic systems in the atmosphere of the Earth we would see the racing, raging currents of the oceans and the remorseless advance and retreat of great ice sheets. On a larger time scale we would observe great lateral movements of the continents, their division and collision and many intricate convolutions, upheavals and tearings of the thin crustal skin. Flooding of large areas of the continents by shallow seas and the subsequent upheaval of this transient sea floor to form new land would be seen again and again. Even more dramatic would be the constant explosive release of spurts of hot gases, with or without incandescent molten rock, from one oceanic or continental volcanic vent after another over the surface of the planet: a kind of widespread safety-valve release or boiling-over going on continuously. The whole effect would be that of a ferociously spitting, bubbling, boiling, convulsive globe with many dramatic and, for periods, continuous upwellings of molten rock along great cracks and rift features cutting right across the continents and oceans. The fact that such a violent and mobile planet can sustain life might well be unbelievable to the crew of a spaceship viewing Earth history in condensed time sequence. To such observers it would certainly appear that any dwellers on the surface of the Earth must face one catastrophe after another and that the chances of survival of a sapient life form from the continuous procession of volcanism, storm, flood, earthquake, continental disruption and bombardment from space were negligible.

Nevertheless, we know that life *has* evolved on Earth, perhaps to some extent because of the ever-present threat of violence and other activity, as well as in spite of these ever-threatening dangers. The evolution of man has been shaped and directed by the forces of nature. We must have survived many great catastrophes over the past few million years – but even if we had a complete knowledge of the hazards that man has been exposed to, this would still be an unrepresentative sample of the potentially catastrophic events which have happened over the 1500 times greater time interval since the Earth itself was born. Our knowledge of events in the Earth's distant past comes from the interpretation of the faint and often circumstantial evidence that is preserved for us in the rocks. The interpretation of this geological evidence is often controversial: indeed, the very basis of our understanding of the evidence contained in rocks has only been widely accepted within the last 100 years. But from this 'record of the rocks', as it was called by the first modern geologist, Edinburgh scientist James Hutton, in the early eighteenth century, geologists today can read Earth history and get some idea of the nature and scale of the evolutionary processes, convolutions, collisions and disasters of the distant past, far beyond that available to man's direct experience. This evidence of the Earth's past contained within the rocks is very incomplete, however, and it becomes more incomplete as the record is traced backwards in time.

Hutton (1726–97) carefully and dispassionately analysed the evidence available in the rocks and countryside around Edinburgh and other places he visited, and from these observations was able to deduce that the slow Earth-changing processes we can observe today – weathering, rain, water, wind and ice causing erosion of the land; the deposition of new rock layers in lakes and seas; volcanism and other forces within the Earth producing or uplifting new land for erosion to destroy – if operative over immense periods of time were sufficient in themselves to explain the record of the past history of the Earth as contained in the accessible rocks of the Earth's Crust. He saw that, in the oldest rocks available for him to study, there was evidence of rock-forming processes in operation very similar to those that could be observed today, and so concluded that the past history of the Earth could not have been so dramatically different from the present. Hutton's concept of the Earth implied an immense time scale: he claimed that he could see 'no vestige of a beginning – no prospect of an end' to geological time. His dictum of 'the present as a key to the past' and particularly its implications regarding the age of the Earth were vigorously attacked by many theologians and scientists. As late as the beginning of the nineteenth century many naturalists still believed that the Earth's landscape had been shaped over a very short period by a few sudden dramatic events. They thought, for example, that all canyons had been carved, oceans filled, and mountains raised by single immense, virtually instantaneous catastrophic events, whereas Hutton saw that the Earth was a regenerative machine and that as fast as continents and individual landscapes were destroyed, other forces were raising new land and new mountains to begin the cycle again.

Many Western scientists of the seventeenth, eighteenth and nineteenth centuries, although influenced by the findings and fieldwork of men like Hutton, were basically religious men and the implications of the great periods of time advocated by the geologists created a conflict between what they saw in nature and what they had been taught to believe by the Bible. The final refutation and eclipse of the biblical view of Earth history began with the publication by Sir

Charles Lyell of his book *Principles of Geology* in 1830. Lyell (1797–1875), a Scot who lived for most of his life in London, was convinced that Hutton was right. He saw nothing supernatural in the simple, straightforward facts of geology, accessible to anyone who travelled in the countryside with an open mind. Lyell's book renewed world-wide interest in Huttonian theory and the impact of his persuasive writings was complemented by the publication in 1859 of the English naturalist Charles Darwin's revolutionary treatise on the *Origin of Species*. It was immediately clear that Darwin's theory of evolution necessitated continuous, irreversible, progressive change in living matter from a remote beginning towards a far distant future: the time scale of Darwinian evolution, like Hutton's Earth history, was extremely long. In the second half of the nineteenth century geologists and Darwinian evolutionists began to talk confidently about the history of the Earth in terms of thousands of millions of years, but they had no way of actually measuring the duration of time past other than by making rather crude guesses based upon such things as the gradual accumulation of salts in the oceans or the total thickness of sediment that had been deposited throughout geological time.

In 1899, this Huttonian/Darwinian view of the Earth received what appeared at first to be a major setback. In that year, the most influential physicist of his day, another Scot, Lord Kelvin (1824–1907), published estimates of the age of the Earth based upon computing the time necessary for the planet to cool from a supposedly initial molten state to its present state of low surface heat flow. Possible differences in the temperature of the Earth at its birth allowed Kelvin's estimates to vary between 200 million years and 20 million years. For a variety of reasons he preferred the lesser figure of 20 million years. Kelvin's time scale, apparently unassailably based upon the evidence of physics, was far too short for the geologists and evolutionists. Were Hutton, Lyell, Darwin and their followers wrong? There was great glee in some quarters. Fortunately this major embarrassing difference of opinion between sciences was almost immediately resolved.

The French scientist Henri Becquerel had discovered radioactivity in 1896. Becquerel's friend Pierre Curie and his Polish-born wife Marie joined him in further investigation of the mysterious radiation that he had discovered was associated with uranium; and in the first years of the twentieth century, work by the Curies and by the Englishman R. J. Strutt (later Baron Rayleigh) showed that the natural spontaneous decay of radioactive elements contained in the rocks of the Earth must provide an enormous additional source of heat not taken into account by Kelvin in his calculations. Medieval alchemists spent their lives in futile attempts to transmute base metals into gold: Becquerel and the Curies showed that natural transmutation of the elements, although not producing gold, was taking place all the time and was producing something of far greater use to man – energy. The further suggestion, that this newly discovered radioactive decay property of certain elements could be used to measure the age of minerals and rocks was first made, in a lecture given at Yale in 1905, by the New Zealander Lord Rutherford, Head of the Cavendish Laboratory at the University of Cambridge. Rutherford's suggestion was immediately followed up by an American, B. B. Boltwood, who showed that it was indeed possible to obtain ages for uranium-bearing minerals separated from rocks of various ages between 1600 million years old and the present day. Thus the science of geochronometry, the measurement of geological time, was born.

Internal heat is the major cause of the Earth's activity. Some part of this heat may be primeval, that is, retained from the gas cloud from which the Earth evolved, but now we know that by far the greatest source of heat energy within the Earth is that derived directly from radioactivity. Natural radioactive decay of uranium, thorium, potassium, aluminium and other elements, supplemented in the earliest times by the decay of transuranium elements no longer present in measurable quantities, provide a continuous, albeit slowly diminishing, heat supply. The Earth is a vast nuclear power station: without radioactive heat it would be a completely dead, inert planet. That similar internal heat is also present in Mars, Venus, Mercury and the Moon is proved by the evidence of volcanism observed on these planets, but on no other inner planetary member of the Solar System has the internal source of energy been sufficient to cause the mobility which is so characteristic of the Earth. The Earth has been likened to a heat engine in unstable equilibrium. For long periods the continuous release of internal heat through scattered volcanism appears to be sufficient to keep the Earth in a state close to stable equilibrium, but important long- and short-period cyclical fluctuations in the heat supply are always disturbing this steady state. When the accumulating internal energy exceeds that being released, then there is an increasing tempo of more violent and catastrophic volcanic and earth-moving activity. Continents are split apart and reunited, volcanoes erupt more frequently, ocean basins are formed and swallowed, mountain chains raised, oceanic deeps appear and new continental Crust is made from a mixture of reworked older Crust, metamor-

phosed sediments and volcanic rocks in great mobile belts bordering
or crossing the more ancient continental nuclei. But as fast as new
mountains and continents are constructed and uplifted, erosion is at
work destroying and levelling the land back to sea level.

How do geologists read this complex story of land construction and
destruction in the record of the rocks? The first necessity is an

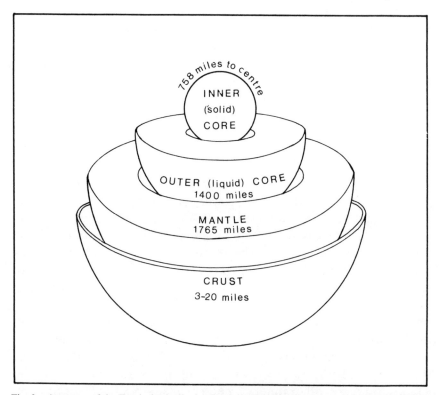

Fig. 3   Anatomy of the Earth: basically the Earth is made up of a series of concentric shells in
much the same way as an onion. The Crust is the outermost shell and is made up of two layers –
the continental Crust, granitic in composition and rich in silicon and aluminium, known as the
Sial, and the oceanic Crust, basaltic in composition and rich in silicon and magnesium and
known as the Sima. The lighter sial (density 2.7 g/cc) is underlain by the more dense (density 2.9
g/cc) sima, which makes up the ocean floor.

The Mantle is the next shell, separated from the Crust by the Mohorovičić discontinuity,
where the rock density suddenly increases from 2.9 to 3.3 g/cc. The Mantle is made up of
peridotite – a dense rock composed principally of iron and magnesium which, near the top of the
uppermost Mantle, is a plastic layer known as the asthenosphere or low velocity zone; seismic
waves travel more slowly in the peridotite here which is near to its melting point. The so-called
lithosphere is made up of the Crust plus the Mantle down as far as the low velocity zone. In the
lower Mantle the density increases from 4.3 to 5.5 g/cc due to higher pressure and closer
packing of the atoms.

The Core is separated from the Mantle by the Gutenburg discontinuity which marks a change
in density from 5.5 to 10.0 g/cc. Here in the outer Core certain seismic waves are not
transmitted, a fact which suggests that it is in a plastic-liquid state. Other seismic waves travel
slower in the liquid outer Core and then speed up in the central Core (density 13.6 g/cc), thus
indicating it is solid and probably similar in composition to some iron-nickel meteorites.

understanding of rocks themselves and a knowledge of how they are formed.

The deepest mines and boreholes are only pin-pricks in the outer-most skin of our planet. From a combination of information from heat-flow measurements, earthquake studies, astronomical observa-tions and the data from remote sensing (instruments and cameras carried by satellites), we can be pretty sure that the Earth has a hot, mainly fluid, nickel-iron rich Core some 4316 miles across. This dense metallic Core is surrounded by a 1765 mile thick rocky carapace or Mantle rich in silicon, oxygen, iron, magnesium, calcium and all the other elements in minor proportions. Only very rarely can we examine the top of this rocky Mantle – one such place is said to be the remote islets of St Paul's Rocks in the South Atlantic – because almost everywhere it is covered by the thin crustal rock skin of the Earth. The Crust of the Earth is usually between 3 and 37 miles thick. It is thickest under high mountain ranges and is several times less thick under the oceans than under the continents. The uppermost part of this Crust is often composed of thin layers of sediments, lava and tuff deposited by wind, water, ice or volcanic activity. Beneath the flat-lying sedimentary-volcanic layers the continents are made of folded and heat-altered rock, often intruded by bodies of cooled and crystalline rock melt. The continental Crust has an overall composi-tion close to that of the rock *granite*; beneath the oceans the Crust has a composition closer to that of the lava rock *basalt*. The base of the Crust is everywhere a sharp discontinuity – called the *Moho* after its discoverer, a Serbian scientist named Mohorovičić – across which the velocity of earthquake waves increases rapidly. The Upper Man-tle, extending some 450 miles below the Moho, has a composition close to the rock *peridotite*, and it is from this region of the Earth that most volcanic lavas are derived and within which most earthquakes originate. Thus, study of this important zone is of great interest to scientists. (When the Mohole Project – the code name for an attempt to drill through the oceanic Crust and examine the Upper Mantle below the Moho – was cancelled by the US Congress in August 1966, this was a bitter disappointment for many Earth scientists and unfavourable comparisons were made with the vast sums of money that were being spent on attempts to put men on the Moon.)

There are three main categories of rock: *igneous, sedimentary* and *metamorphic*, classified according to how they are formed. *Igneous* rocks are mineral aggregates produced by the relatively slow crystal-lization of cooling rock melts. Very occasionally, a rock melt may be so rapidly cooled or quenched that a natural rock glass (e.g. obsidian) is produced. Molten rock may cool and consolidate within the Earth's Crust as *intrusive* bodies, such as granite batholiths, dolerite dykes and sills; or it may break through to the surface and *extrude* to form either lava flows and lava volcanoes or be erupted explosively as ashes, agglomerates and volcanic breccias to form ash- or cinder-cones and ash-flow sheets.

Prolonged weathering breaks down all surface rocks into easily eroded fragments and material that goes into solution. This erosion debris is transported by wind, water and ice and deposited in low-lying places and in the lakes, seas, and oceans of the world as layers of sediment. Some *sedimentary* rocks (called clastic sediments) are made entirely of erosion debris (e.g. some sandstones and shales); some are made of erosion debris cemented together by precipitates from the material in solution in surface waters; others are entirely chemical in origin (e.g. some limestones, rock salt); and others again are rich in organic or organically precipitated matter (e.g. coal, coral limestone).

Deep burial of rocks causes gradual mineralogical changes to occur in response to increasing pressure and temperature. In mountain belts deep burial may be accompanied by intense deformation, folding and faulting: rocks so altered are known as *metamorphic* rocks (e.g. gneiss, schist, marble, slate, hornfels).

When outcrops of rock are examined in the field (e.g. in rocky crags, stream beds, ravines or sea cliffs) diagnostic features can be observed which enable a trained geologist to deduce how they were formed. Careful study of rocks formed today – in modern volcanic eruptions, by modern rivers, in modern lakes, seas and oceans and also by the simulation of the processes involved in the laboratory – has revealed the special characteristics of various different modes of formation. The presence of these same characteristics in ancient rocks means that the ancient rocks are likely to have been formed in a similar manner.

As an example of how we use this technique, let us examine some of the characteristic features of wind-blown (aeolian) sand dunes found in modern tropical desert areas. Each dune has a long flat windward slope rising to a sharp crest which separates it from a short, steeper and gently concave leeward or downwind slope. Strong winds blowing on the windward face pluck out sand grains and move them upwards and over the crest of the dune. Once blown over the crest, the sand falls into a wind shadow and comes to rest at its natural angle of repose – about 30° to 35° for dry sand – each skin of sand being buried by succeeding layers as the wind gusts over the dune. Barchan

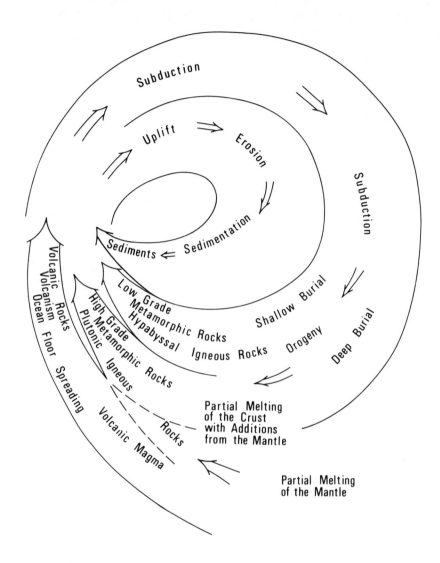

Fig. 4 The rock cycle, illustrating the relationship between igneous, metamorphic and sedimentary rocks. The main processes involved are volcanism and uplift producing new land areas, erosion of this land and subsequent sedimentation to make new sedimentary rocks, deep burial and mountain building converting all pre-existing rocks into metamorphic rocks, and partial melting at the greatest depths to form rock melts (called *magmas*) that move upwards and consolidate as new igneous rocks. The rock cycle has been continuous since very early in the Earth's history. The principal activity of the continental rock cycle is entirely within the thickness of the Crust, but repeated additions of partial melt and other fluids from the Mantle have greatly increased its volume over geological time. The heavier oceanic Crust, mostly volcanic in origin with a thin skin of deep water sediments, is normally recycled within the Mantle.

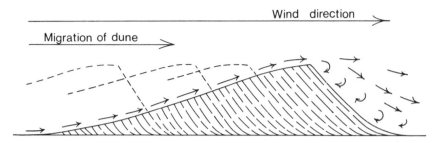

Fig. 5 Cross-section of an aeolian sand dune: the dune migrates forward in the direction of the wind by a process of erosion on the windward slope blowing sand over the crest, where it falls into the wind shadow on the leeward side and accumulates in layers at its natural angle of rest.

and transverse dunes move forward by this process of erosion on the upwind side followed by deposition on the leeward side. If a trench section is cut through such a dune from the upwind to the downwind side, a sequence of fine, gently concave sloping lines will be seen on the sides of the trench which represents bedding planes (i.e. the top and bottom surfaces of simultaneously deposited layers of sand). These bedding planes are truncated sharply by erosion on the upwind face of the dune. On the downwind side they represent successive former surfaces of the dune as it progressed slowly forward. When examined with a microscope or hand lens, many individual sand grains in the dune show characteristic polished surfaces due to pro- longed sand blasting in desert conditions. In the downwind areas of extensive desert basins, where many dunes may have accumulated, building up one on the other, more complex-inter-dune bedding relationships can be revealed by trenching. Again, in the more deeply buried parts of such a pile of modern dunes a red iron-oxide, sili- ceous and/or calcareous cement derived from ground water may be observed in the process of coating the surface of individual sand grains and filling the cavities between grains. A comparative study of dune formation in numerous modern deserts has given geologists a very comprehensive understanding of the characteristic features of rock formed in this manner.

In the English Midlands, around such towns as Kidderminster and Bridgnorth, where the climate today is temperate and maritime, there are many road cuttings in a red rock that is known to be about 230 million years old. Close examination shows that this rock layer is a sandstone containing numerous wind-polished sand grains partly coated and cemented together with red iron oxides, calcite and chal- cedony. In addition, many outcrops of the rock layer reveal exactly the same truncated, gently curving bedding planes that are seen in sections of modern dunes. Thus, using Hutton's dictum 'the present as a key to the past', we can deduce that this red rock layer exposed in the English Midlands is an ancient dune complex that accumulated some 230 million years ago in a tropical desert. This deduction tells us a number of things about the past geological history of Britain. It is clear that at that time, the English Midlands were part of a large continent lying close to the Equator and that the geography of England was dominated by extensive intermontane desert basins, similar in many ways to the desert basins of the modern world, where dune complexes are being formed today. Eastwards from England, the rocks deposited in this ancient desert can be traced across the North Sea through Germany and Poland to Russia. In America they can be followed from New York to Texas. We know now that 230 million years ago the Atlantic Ocean did not exist. From Texas to Russia the desert associated with the Tropic of Cancer lay in a wide belt across a single great continent. A journey from the English Midlands to New York, at that time, would have been like traversing the present Sahara desert from Khartoum to Timbuktu.

In an exactly analogous way careful studies of all aspects of modern volcanism, metamorphism, sedimentation, mountain building and other earth processes enable us to learn how to interpret the numerous, sometimes faint, but always diagnostically characteristic 'clues' frozen into or otherwise shown by ancient rocks which enable us to deduce their mode of formation and, occasionally, to follow the progress of later events in their geological histories.

The 'clues' present within rocks are many and varied. They may be, for example, such things as fossilized rainprints, mudcracks, current bedding, worm burrows, animal trails, footprints, eggs, excreta, sea cliffs, shingle beaches, coral reefs, lava tubes, river point-bars or even whole fossil mountains buried beneath later rock layers for millions of years! On a smaller scale, much can be deduced about ancient mountain building events from the style of any folding present and from the orientation of the joints, faults, cleavage and schistocity planes cutting the rocks. Small differences in composition, the interrelationships of component minerals as seen beneath the microscope, the composition of late cross-cutting mineral veins; the populations of microscopic animals and plants enclosed by rocks: all of these and much other evidence contained by rock may be of use. The trained geologist gathers these clues and uses them to interpret and understand the succession of past environments in exactly the same way that Conan Doyle's famous fictional hero, Sherlock Holmes, gathered and used obscure facts to help solve problems in criminal detection. The intellectual satisfaction obtained from interpretative geology can be easily as great as that obtained from the solution of a sequence of complex and rather tricky murder mysteries!

The next requirement, once we can correctly interpret the clues contained within individual rocks, is to learn how to put together the evidence from many different rock outcrops into a single historical sequence. The science of sequential Earth history is called stratigraphy. A canal builder called William Smith is generally regarded as the 'father of stratigraphy' because it was he who, in the eighteenth century, first clearly understood that the successive layers of rock seen in an east–west traverse across England from Bristol to London – each a different sedimentary rock containing a different assemblage of fossil shells, plants, fish or animal remains – represented successive pages of the local Earth history. In areas like eastern England, where the sequence of sedimentary rock layers are only gently tilted towards the North Sea basin, it is clear that the history they contain is to be read from the lowest layer upwards. This simple rule is called the 'Law of Superposition': it draws attention to the fact that younger rocks are normally deposited on top of older rocks.

Earth history may be much more complex than it is in eastern England, however, as the simple succession of sedimentary rock layers may be interrupted at any time by such things as volcanism, uplift, subsidence, folding, faulting or metamorphism. When an episode of folding and uplift (mountain building) brings to a close a lengthy period of sediment accumulation in any area, not only will there be a sharp break in the historical record preserved in the rocks but deformation and metamorphism may blur or partly destroy the story already preserved. In some episodes of mountain building, rocks may be overturned or even piled up in slices one on the other in incorrect age sequence. Much careful fieldwork is necessary to unravel these complex areas. Usually the only record we have of the long period when an area was uplifted, say as part of a continent or mountain range, is a final sharp cross-cutting erosion surface (a land surface or plain of marine encroachment) truncating the old rock layers and upon which the next, possibly much younger, sediments to be deposited in the area now lie. These important breaks in the geological record are called *unconformities*. Some unconformities represent immense intervals of geological time, the record of which for that area is irrecoverably lost. In order that the geological sequence and history of an area can be properly studied, it is usually an essential prerequisite that a detailed geological (or rock) map of the area be prepared. Geological maps are essential for the evaluation of mineral resources, and since the Geological Survey of Great Britain was founded by Henry de la Beche in 1835, official rock map making survey offices have been opened by almost every country in the world. Today the Geological Surveys of the USA, the USSR and China each employ thousands of geologists.

The history of the Earth as a whole, therefore, must be put together by a process of equating and placing in their correct order innumerable incomplete fragmentary sequences of events derived from the mapping and study of local rock successions throughout the world. A prime requirement in this process is an ability to date rocks and so equate events between distant areas. In relatively young rocks, i.e. those less than 600 million years old, the fossil record of the evolution of life on Earth has given us a relative time yardstick against which different rocks can be compared. Thus when we find rocks containing fossil dinosaurs we know them to be older than those containing fossil man. Similarly, we know that with one exception the fossil invertebrate animals called trilobites are found only in rocks between 600 and 365 million years old. Not all fossils are time-diagnostic, however: the small horny shellfish *Lingula*, for example, has lived unchanged in the world's oceans for 580 million years. To be of maximum stratigraphical value, a fossil species must have been geographically widespread and have a restricted time range (i.e. because it belonged to a rapidly evolving stock and/or became extinct soon after its first appearance). Some fossil species of planktonic oceanic foramenifera have been ideal in this respect, and study of the fossil forameniferal populations of the rocks encountered in boreholes has proved invaluable to oil men, enabling them to correlate rock horizons from borehole to borehole over great distances.

The sedimentary rock sequences that accumulated on various parts of the European continent over the past 580 million years were arranged into a classified sequence by nineteenth-century geologists and the various divisions were given names derived from the localities where the rocks of that age were first studied in detail. Thus, for example, the three oldest divisions of this sequence are known as the Cambrian, Ordovician and Silurian after tribes which inhabited Wales in pre-Roman times, the Permian was named after the city of Perm (Molotov) in the Urals and the Jurassic is named after the Jura Mountains to the north-west of the Alps. Table 2.1 lists the major divisions of geological time so derived. From its beginnings in Britain, North America and Europe, the science of stratigraphy was extended rapidly throughout the remainder of the world. Correlation of rocks by means of fossils is not everywhere unambiguous, however, and

Table 2.1   The geological timescale

Millions of years past	Period	Era	Events
0	Quarternary ∿1.8 n.y.		The last ice age.
	Neogene		Evolution of man.
24 ± 1		Cenozoic	Uplift of the Alps, Rockies and Himalayas.
	Palaeogene		Continental drift advanced.
			The age of mammals.
67 ± 1			
100	Cretaceous		Giant reptiles become extinct. Ammonites die out. Laramide mountains. Chalk seas in Europe.
137 ± 5		Mesozoic	Flowering plants expand rapidly. Extensive shallow seas flood the continents.
	Jurassic		Age of the dinosaurs. Flood basalts in Africa and South America.
200			Arid continental interior of Gondwanaland
210 ± 10	Triassic		
237 ± 10			
	Permian		First mammals. Appalachian mountains uplifted.
285 ± 10			Southern Hemisphere ice age.
300		Upper Palaeozoic	Variscan mountains uplifted. Pennsylvanian coal swamps.
	Carboniferous		Amphibians and reptiles reach large size. Shallow seas deposit shell and coral-rich limestones.
367 ± 10			Arid intermontane basins in north-west Europe. Plants spread on lands.
	Devonian		Caledonian mountains.
400			
415 ± 10	Silurian		
445 ± 10			First air breathers. Corals and trilobites abundant. Early Caledonian folding.
	Ordovician		Graptolites abundant in seas.
500		Lower Palaeozoic	First ostracods and conadonts, seaweeds and nautiloids.
510 ± 10			First graptolites. Algae and brachiopods abundant.
	Cambrian		Trilobites dominant. Extensive seas.
580 ± 15			
600	Precambrian ↓		

cannot be used to correlate Precambrian rocks (i.e. those older than 600 million years). As Precambrian time represents nearly seven-eighths of the Earth's history this was a very serious drawback and little progress could be made towards unravelling the early history of the Earth before the radiometric dating of rocks became possible. A considerable volume of reliable radiometric dating has been undertaken, however, since the late 1950s and this has enabled calibration of the fossil-based time scale for the Cambrian to Recent rocks in years, and an accurate estimation of the total age of the Earth as well as precise dating and worldwide correlation of many of the significant events in the Precambrian history of our planet.

How do we actually measure the age of rocks? As we have seen, nineteenth-century geologists developed the techniques and experience necessary to enable them to arrange the rocks observed in nature into their correct relative age order. Once this has been done for large parts of the Earth, a study of rock forming processes allowed them to make informed guesses regarding correlations between different continents and the length of time required for the Earth's obviously very extensive history to be enacted. Nevertheless, it was not until the beginning of the twentieth century, with the work of the late Arthur Holmes, Emeritus Professor of Geology at the University of Edinburgh, that age measurement, i.e. the actual dating of ancient rocks in years, became possible. Holmes' name is most closely associated with the increasingly successful use of the radioactive decay of elements to date the earth and provide a calibrated geological time scale; today scientists can use one of several so-called 'radiometric clocks' to date rocks and their constituent minerals. These clocks depend upon the natural spontaneous radioactivity decay of isotopes of certain elements (parent isotopes) present within rocks and minerals, a decay which transmutes them into isotopes of other elements (daughter isotopes).

Table 2.2 Major radioisotopic dating methods

Method	Half-life* of radioisotopic isotope	Effective dating range	Materials that can be dated
Uranium–lead		From 10×10^6 yr	Zircon, uraninite, pitch-
Thorium–lead		to the age	blende, monazite (some
$^{238}U/^{206}Pb$	4.51×10^9 yr	of the Earth	whole rocks).
$^{235}U/^{207}Pb$	0.71×10^9 yr		
$^{232}Th/^{208}Pb$	13.9×10^9 yr		
($^{207}Pb/^{206}Pb$)			

Method	Half-life* of radioisotopic isotope	Effective dating range	Materials that can be dated
Potassium–argon $^{40}K/^{40}Ar$ (two methods: total degassing and $^{40}Ar/^{39}Ar$ age spectrum analysis)	$1 \cdot 31 \times 10^9$ yr	From 10^5 yr to the age of the Earth	Any potassium-bearing mineral or rock.
Rubidium–strontium $^{87}Rb/^{87}Sr$ (two methods: mineral ages and isochron analysis)	$48 \cdot 5 \times 10^9$ yr	From 10×10^6 yr to the age of the Earth	Rubidium-rich minerals or rocks such as muscovite, biotite, microcline, granite-gneiss, etc.
^{14}C	5730 ± 30 yr	0–40 000 yr	Wood, charcoal, peat, grain, tissue, charred bone, cloth, shells, tufa, groundwater.
Fission track ^{238}U	Spontaneous fission rate $c.\ 10^{-16}$ yr	From 6 months to the age of the Earth	Glass, apatite, sphene, zircon, epidote, allanite, hornblende, garnet, pyroxene, feldspar, mica, etc.

* The half-life of a radioisotope is the time in which one half of a given amount of the isotope will have decayed. It is related to the decay constant, λ by the formula:

$$\text{half-life} = \frac{0.693}{\lambda}$$

Today, in the late 1970s, we estimate that the Earth is 4600 million years old. This conclusion is based on a number of mutually supporting lines of evidence: (i) the oldest terrestrial rocks found so far are a sequence of metamorphosed volcanic lavas and sediments exposed inland from Godhaab in West Greenland at a place called Isua. Radiometric dating indicates that they accumulated around 3800 million years ago. The presence of normal clastic sedimentary rocks in this extremely ancient rock sequence confirms the pre-existence of even older primitive crust supporting land areas and oceans and with the geological processes of precipitation, weathering and erosion in progress. Thus the Earth itself must be considerably older than 3800 million years; (ii) age determinations on meteorites yield ages close to 4600 million years and meteorites can be regarded as debris left over from the earliest stage in the evolution of the planets; (iii) age determinations of rock samples collected from the Moon yield ages close to 4600 million years for the formation of the oldest rocks found; (iv) the lead isotopes ^{206}Pb, ^{207}Pb and ^{208}Pb are end-products of the uranium-thorium radioactive decay series. The age of the Earth-Moon-meteorite system can be estimated from consideration of the gradual evolution of lead isotopes with time. Such an analysis suggests that lead with a primeval isotopic composition was included in this closed system with uranium and thorium around 4600 million years ago.

As we said at the beginning of this chapter, 4600 million years is an incredibly long time span for a human being to comprehend. In order to make the Earth's history more easily understandable, suppose we call 1000 million years one *earth-day*. On this time scale, we are today in the fifth day of the Earth's life: four days, fourteen hours and twenty-four minutes from the hour of its birth! The oldest dated rocks, at Isua in West Greenland, were formed around 7.12 p.m. on the evening of the first earth-day. Primitive life was certainly present in the early oceans during earth-day two, but the great evolutionary explosion that has produced the multiplicity of modern life forms did not begin until after midnight at the end of earth-day four. Days one to four represent Precambrian time, the record of which is contained in the predominantly unfossiliferous crystalline rocks of the continental basements. In the early hours of earth-day five, life in many forms was proliferating in the oceans. Around dawn on that day colonization of the land began in earnest: a memory of the great forests of tree-ferns and other swamp plants that covered the deltaic margins of the low lying northern continents by around 7 a.m. has been preserved in fossil form as the Upper Carboniferous (Pennsylvanian) coalfields of Europe and North America. Those great reptiles, the dinosaurs, ruled the Earth throughout the whole of the rest of the morning of day five, only becoming extinct just before 1 p.m. The early afternoon of earth-day five (from about 1 p.m. to the present time at 2.24 p.m.) has seen the great expansion of birds and mammals. Early man appeared about 2.17 p.m. Our own species (*Homo sapiens*) has only been around on Earth for the last 7.2 seconds!

Nowhere on Earth can man better appreciate his own insignificance than at the Grand Canyon of the Colorado River in Arizona (Plate 6). Standing on the southern edge of the Canyon at Canyon View, more than 50 miles of a dramatic mile-deep section through the Earth's outer Crust can be seen in one breathtaking panoramic view. This enormous 4 to 18 mile wide gulley has been excavated by the Colorado River over the past 10 million years (i.e. in the most recent 15 minutes of earth-day five) as this part of America has been slowly uplifted and uparched. The Canyon is easily big enough for the burial of a single great coffin containing the entire human race! The geological section exposed covers more than 2000 million years (two earth-days), representing nearly half of the Earth's history.

Within the main canyon the Colorado River flows in the narrow inner or Granite Gorge. The walls of Granite Gorge expose a very thick sequence of lavas, tuffs and sandstones that show the characteristic features of rocks formed in those mobile belts of the Earth's Crust within which mountains are made. At least 25,000 feet, probably much more, of these volcanic and sedimentary rocks accumulated before they were intensely folded, metamorphosed and intruded by great volumes of molten granite in an important ancient fold mountain building event that occurred in the latter part of earth-day three. For many millions of years these great mountains of crumpled metamorphic rocks were a dominant feature of the old continent of which they formed a part. There is evidence that rejuvenation of the original mountain belt, with accompanying intrusion of granite and dolerite dykes may have occurred on more than one occasion. Eventually, however, erosion wore them completely down, reducing the whole area to a flat-lying peneplain. To do this, several miles of rock had to be eroded and the debris transported away to fill younger continental and marine sedimentary basins that had been initiated elsewhere.

In late Precambrian times (the afternoon and evening of earth-day four) the sea returned and a further 12,000 feet of sediments were laid down in the area of the present-day Grand Canyon. These rocks (of the so-called 'Wedge Series') are separated from the underlying crystalline basement by a great unconformity representing the time interval during which uplift and erosion were the dominant earth processes at this location. The Wedge Series includes sandstones, conglomerates, shales and limestones. Much of the succession is a distinctive red or vermilion colour: certain of the limestones contain structures which are interpreted as indicating that they were algal reefs built by the activity of very primitive plants. Then, late in Precambrian times (a short time before midnight at the end of earth-day four) the area was uplifted once more by another period of earth movements. The upper part of the crust was broken and sliced through to the surface by a group of great faults. The movements on these faults would have caused a great number of earthquakes, some catastrophically violent. Between the faults the intervening ground was disturbed, and tilted in a series of giant blocks. Inevitably, as soon as these block mountains were formed, erosion began to destroy them, eventually (by midnight on earth-day four) reducing them to another low-lying peneplain, one that cut right through the Wedge Series in places to expose the underlying crystalline basement. Thus, the Wedge Series is only preserved and seen today as tilted, wedge-shaped blocks faulted against the crystalline basement at their thick ends and cut off above by the second great unconformity.

In the Grand Canyon, the great unconformity above the Wedge Series marks the end of Precambrian times. Overlying this uncon-formity are several thousand feet of flat-lying rocks representing the history of earth-day five – our day. There are about 4000 feet of these rocks up to the canyon rim and the story is continued through a further 10,000 feet of sediments in the scarps of northern Utah. All of these rocks have been gently arched across the Colorado Plateau. At the beginning of earth-day five the peneplained surface of the late Precambrian continent sank back slowly beneath the sea. The advancing waves of the Cambrian sea planed off the surface to produce the actual surface of unconformity visible today. The debris so formed was spread out on the sea floor as shingle and sand. As the sea deepened the dominant sediment being deposited at any one point changed from sand to mud. Eventually the dominant sediment was limestone. These rock layers contain numerous examples of the same fossil arthropods called trilobites that were found in the original Cambrian rocks of Wales and thus we know them to be of Cambrian age.

Rocks representing the remainder of Lower Palaeozoic time are not seen in the Grand Canyon. It is probable that the area remained close to sea level for about 100 million years during the Ordovician and Silurian Periods (from about 2.24 a.m. to about 4.45 a.m. on earth-day five) and that little, if any, rock belonging to these systems was deposited. The next event was a gentle uplift above sea level. Erosion following this uplift stripped off any thin layers of Ordovician or Silurian rocks that may have been deposited and cut shallow river valleys into the underlying Cambrian rocks. River-deposited sand-stones and limestones of this period contain the fossilized remains of the early armoured freshwater fish and other animals that are typical of Devonian times. These fish-bearing rocks are only found as thin discontinuous lenses beneath a great red-stained cliff of limestone, the Redwall Limestone, some 550 feet thick, that follows and is one of the most prominent features of the Grand Canyon, occurring about midway in its walls.

The Redwall Limestone is a pure marine limestone very rich in the remains of sea life of many forms. It tells us that at around dawn on

earth-day five the area of the Grand Canyon was inundated by the
Lower Carboniferous (Mississippian) ocean. Towards the end of
Carboniferous times (say about 7.15 a.m.) this sea retreated and
desert conditions spread over the new land area so formed. Shales of
this age contain land animal trails, fossil ferns and insect wings.
Above them are steep cliffs in a prominent white aeolian sandstone,
the Coconino, complexly cross-bedded and unfossiliferous again but
for animal tracks. It is clear that these late Carboniferous and Permian rocks were deposited in a desert by wandering streams and in
temporary lakes: the white sandstone is a dune complex. Towards the
end of Permian times (around 8.15 a.m. on earth-day five) the land
began to sink beneath the sea once more. At first the shoreline
oscillated back and forth, but the final Permian sediment, outcrop-
ping on the rim of the Canyon, is a marine limestone rich in sea-shells,
corals and sharks' teeth. This late Permian Kaibab Limestone brings
to an end the Palaeozoic Era (the era of early life) at the Grand
Canyon (it was then about 8.53 a.m. on earth-day five).

The story of the fifth day of the Earth's history can best be continued away from the Grand Canyon, in northern Utah, at Red Butte
and Cedar Mountain and in the Painted Desert and Zion Canyon
region. From these rocks it can be deduced that at the beginning of
the Mesozoic Era (the middle era of abundant life on Earth) the sea
again retreated and Arizona and Utah once more became part of a
great desert and semi-arid continent. The red conglomerates, sandstones and shales of the Triassic contain fossilized forests at places.
Dinosaur footprints are found in river deposits of early Jurassic age
(around 8.40 a.m.). The most striking rock of Jurassic age is the 1000
foot-thick Navajo Sandstone that forms the White Cliffs escarpment
in Utah, the walls of Lake Powell above the Glen Canyon Dam, the
walls of the Escalante Canyon and the dome of Navajo Mountain in
northern Arizona. The Vermilion and White Cliffs cut in this rock
and forming the walls of Zion Canyon and Echo Cliffs north-east of
Navajo Point on the Grand Canyon, have been aptly called 'cliffs of
fossil sand dunes'. To the west of the Grand Canyon area, a desert
and semi-arid continental environment persisted right through the
Mesozoic Era to the end of Cretaceous times some 67 million years
ago (i.e. from about 8.53 a.m. until about 12.45 p.m. on earth-day
five). Further east in Colorado and New Mexico marine conditions
existed. In the intervening area the sea advanced and retreated so
that the rock record of the Cretaceous consists of alternations of
marine sediments rich in shells and other fossils; coal deposits
indicate the existence of low-lying coastal plains and swamps with

Fig. 6 *Tyrannosaurus rex*, the largest carnivorous dinosaur known: fast moving and 50 feet from head to tail, this monster had a mouth full of murderous 8 in. teeth. It would have been a fearful hazard to any creature as puny as man.

luxuriant vegetation (the habitat of numerous large dinosaurs) and river-borne conglomerates, sandstones and shales derived from higher land in the west. Around 12.45 p.m. on the fifth earth-day the 'age of dinosaurs' gave way to the 'age of mammals'. The Grand Canyon area has been above sea level during the whole of the Cenozoic Era (the era of recent life). The rock record is sparse: it does include, however, the thick fresh water lake deposits of Bryce Canyon and the volcanic lavas of the San Francisco Mountains. Much of the history of the Grand Canyon area in Cenozoic times is a story of erosion: the production of the great escarpments and buttes, the levelling of extensive peneplains and, latterly, warping, uparching and the excavation of the great canyon by the Colorado River. Today, at 2.24 p.m. on the fifth earth-day, the Grand Canyon is one of the great natural wonders of the world, but it is, of course, ephemeral in geological terms. Before long, erosion will cut away much of the present-day uplands of Arizona and Utah and, inevitably, within a few earth-hours, the area will be peneplained once more and the time will come for it to sink beneath the ocean again as it has done so often in the past.

This review of the geological history of the Grand Canyon area of northern Arizona and southern Utah reveals that over much of its past this area would not have been habitable by man. Two thousand million years ago it was a deep oceanic trough; from the volcano-sedimentary fill of this trough a belt of high fold mountains was formed. These vegetation-free mountains in the process of rapid decay would not have been as habitable as the Moon is now, and the episodes of rejuvenation they suffered would certainly have been accompanied by occasional catastrophic earthquakes and by extensive volcanism. In late Precambrian times the ocean returned to flood the area and after a brief interval during which the area was uplifted as block mountains it returned again, probably remaining throughout Lower Palaeozoic times. A semi-arid land with fish-rich rivers and lakes that might have supported man if he could adapt to the low oxygen content of the air did exist for a short time in the Devonian Period, but this short-lived land area was once more inundated by the ocean during the Lower Carboniferous. Then followed uplift and a very lengthy period when the Grand Canyon area of North America was part of a huge desert and semi-desert continent: once again man might have been able to live a rather poor nomadic life under such conditions. Even this land was rather impermanent, being inundated by the sea on a number of occasions. Nevertheless, plant and animal life was very abundant at certain times along the coastal plains and man could probably have made good use of these parts of the continent. He would have had to avoid the attentions of the great dinosaurs, however, who were the dominant creatures of these times and were most numerous in the fertile coastal zone! Since the beginning of the Cenozoic Era, 67 million years ago, man could probably have lived reasonably satisfactorily in the Grand Canyon area for most of the time. Occasionally, the area may have been too arid and at other times subjected to devastating volcanism, but for most of the period the environment would not have been too different from that of today. Of the future, one thing is certain: the ocean will return to Utah and Arizona on more than one occasion, driving man and his fellow land animals, if any are still present in the area, inexorably before it.

In the preceding pages we have briefly reviewed the development of the science of geology. We have seen that geologists can date rocks using the radioactive decay of certain elements contained within them and how, by the careful collection and examination of various 'clues', they can interpret the succession of past environments represented by rocks. We have learned that the Earth is 4600 million years old and looked in some detail at the history of Arizona and Utah over the second half of the Earth's life, seeing how it is possible for trained geologists to 'read the records of the rocks' and to examine the past for evidence of events in the history of the Earth that, if they were to be repeated, would be catastrophic disasters to civilized man and his works. Even a quick and quite cursory review of the history of the Earth as recorded in the rocks of the outer Crust reveals an endless succession of such potentially disastrous events.

Very early in the Earth's history the protoplanet would have been a great ever-growing ball of rock, dust and gas orbiting the Sun within its sector of the primitive solar nebula, continuously sweeping up the remaining matter within its gravitational reach rather in the way that a giant snowball grows as it rolls downhill. As the protoplanet grew in size and contracted under gravity its internal temperature would have risen sufficiently for partial melting, gravitational segregation and volcanic outgassing of its rocky substance to have begun. Many of the lighter gases initially collected by the protoplanet or released into its atmosphere by various outgassing processes would have been rapidly blown away by the solar wind. The early atmosphere would have consisted very largely of an unbreathable mixture of hydrogen, water vapour, carbon dioxide and nitrogen. During the later stages of initial growth, the surface layers of the protoplanet would have been subjected to continuous bombardment by smaller, less successful, condensations from the primitive solar nebula. The fall of these large and small microplanets, or *planetisimals* as they were called by the American geologist Chamberlin, would have reduced the surface layer of Earth to a saturation-cratered dust bowl and the heat released by multiple impact may even have resulted in extensive melting of the surface layers. We have no idea how long this early chapter in the Earth's history may have lasted. Estimates vary between 100 years and 100 million years. It is certain, however, that until this violent period of repeated impact, explosion, melting and volcanism was completed, very little of the water vapour pouring into the atmosphere as volcanic gas clouds could have begun condensing to form even temporary seas, let alone initiate the first oceans. Man certainly could not survive on any planet in this stage in its development. The incessant rain of bombarding planetisimals and meteorites was still very high during the first 580 million years of the life of the planet: probably between 1000 to 10,000 times the rate at present experienced by the inner terrestrial planets.

Around 4000 million years ago, while the cratering rate was falling rapidly, volcanism became a dominant feature of the Earth. It is probable that early in its life the Earth experienced a few enormous impacts such as those that produced the maria on the Moon and that the gigantic impact craters up to 2500 miles across so formed became filled with floods of basalt and other lavas even more rich in magnesium and iron. On the Moon these early giant craters are still clearly visible, but, because the Earth's Crust is in constant mobility and partly as a result of the erosion and later deep burial so often suffered by ancient rocks on Earth, they are very difficult to identify here. Within the oldest continental terrains, deep infolds of highly metamorphosed volcano-sedimentary rocks form part of the continental nucleii. Recently it has been suggested that some of these infolds represent impact crater fill, squashed, folded and metamorphosed during subsequent periods of crustal mobility.

The Earth's atmosphere and oceans accumulated largely as one of the end products of volcanism. It was in this period, between 4000 and 3000 million years ago, that the oceans first became extensive and a land and sea environment, albeit dominated by volcanism and extraterrestrial bombardment on a vast scale, became established – an environment that would have been recognizable in terms of the modern geologist's view of the Earth. With very little oxygen, the atmosphere would still have been unbreathable, however, and, even discounting this, survival of human life from the ever-present hazards of frequent bombardment and widespread volcanism would have been virtually impossible. The continental Crust was thin in these Archaean times: it crumbled more readily and extensively and was more frequently injected by great volumes of liquid granite than later in its history. Elsewhere it was readily torn apart and injected by great wedges of basalt liquid (now seen as dolerite dyke swarms) derived from partial melting of the Upper Mantle. These dyke swarms often fed vast outpourings of flood basalt lavas on the surface.

During the period between 3000 and 2000 million years ago the Earth's surface was less frequently bombarded by meteorites and planetisimals, the rate of cratering being on average no more than about five times the present rate. Continents and oceans existed and in the oceans primitive life forms were proliferating. In the first half of the period great mountain belts appeared in all of the continents accompanied by extensive deformation and folding of older rocks, earthquakes, volcanism, metamorphism and intrusion of granite. Erosion of these mountainous areas poured thick fills of sediment into numerous intermontane basins and into the sea. Further dyke swarms of dolerite and large intrusions of a rather special rock-type also known from about this time on the Moon (*anorthosite*) were typical of the later half of the period.

We have already examined the last half of the Earth's history in some detail, as it is recorded in the rocks of the Grand Canyon. From evidence gathered in a similar way elsewhere and on other continents we know that soon after 2000 million years ago yet another world-wide period of enhanced mountain building corrugated and uplifted great belts crossing each of the continents. Again, as soon as they began to form these mountains were subjected to erosion and decay, and this process of worldwide mountain building and subsequent planation by erosion was repeated twice more in the Earth's history, around 1000 and 500 million years ago respectively. Less widespread *orogenic* (i.e. mountain building) events of similar kind to those mentioned above have occurred locally at other times throughout the Earth's history.

Evidence of many great earthshock catastrophes is hidden in the Precambrian record of the rocks. Every large fault against which crustal rocks are displaced must have been the source of numerous earthquakes when it was active. The thick layers of lava, tuff and *ignimbrite* (welded incandescent pyroclast-flow tuff) exposed in the rock sequences tell of violent and prolonged volcanism. The sedimentary record is one of numerous advances and retreats of the ocean across the land. The repeated erosion of high mountains cannot have been accomplished without many a storm, flash flood, avalanche, ice-fall, rock burst, earthslide or similar disaster. In addition, there was the constant hazard of extraterrestrial bombardment, the rate of which did not fall to about the present rate until after 2000 million years ago.

Some idea of the devastation caused by such collisions between the Earth and asteroid-sized bodies approaching us from space can be obtained by looking at a geological map of the Johannesburg district of South Africa. Many geologists regard the ring structures visible within the ancient rocks north and south of that city as fossil scars left in the Earth's Crust from a multiple planetisimal impact event that occurred some 2100 million years ago. South of Johannesburg, the Vredefort ring, about 50 miles across, is surrounded by a much larger annular zone of related crustal disturbance. To the north, three similar overlapping ring complexes each more than twice as large as the Vredefort ring combine to form the Bushveld structure. In all, a roughly triangular area of continental Crust around Johannesburg each side of which was 300 miles long was severely disturbed and punctured by the four contemporaneous impacts. Today we can only examine the deformed Crust and the rocks formed from melts that were intruded into it far beneath the former craters. The old ground surface and the craters themselves have long since been removed by erosion. We can guess, however, that the surface effects of this multiple impact must have been tremendous, with destruction and burial of the ground beneath layers of ejected rock fragments extending for thousands of miles away from the group of giant craters. A more detailed discussion of this very disturbing but fortunately very rare kind of earthshock will be found in Chapter 7.

Only in the last 1000 million years of the Earth's history has the environment approached one in which human life would have been at all possible. In fact, it was not until after life spread prolifically onto the land during the Upper Palaeozoic that conditions arose in which human life could have been anything like pleasant on our planet. Thus, the period of the Earth's story of most interest to us is the last 400 million years. It is the events and disasters of these years that are most relevant to the present situation of man and his future. During

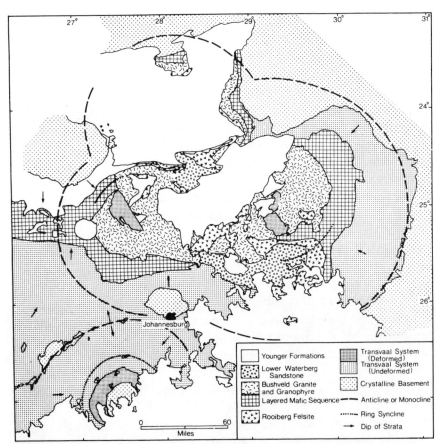

Fig. 7 Geological map of the country around Johannesburg, South Africa. The Vredefort Dome and Bushveld structures have been interpreted as the scars of a multiple impact by four great rocky missiles from space around 2100 million years ago. Today we can only see a record of the event in rocks that were then far below the surface. The giant craters that must have formed on impact and the debris covered ground surface that must have extended for a thousand miles or more from the vicinity of Johannesburg have long since been eroded away.

the first 100 million years or so of this period we now know that a great single supercontinent came into being, with the joins between various former separate crustal fragments being welded by mountain belts. Then, since 250 million years ago, the Earth's history has been very largely one of the subsequent disruption and dismemberment of this supercontinent. The mechanism of this, most probably oft-repeated process of continent growth and subsequent disruption, *the dance of the continents*, has only been properly understood by Earth scientists very recently, but it is of such importance in understanding the causes of earthquakes, volcanism, mountain building and climatic change that Chapter 3 will be entirely devoted to this subject. Detailed examination of the rock record for the last 400 million years confirms that the principal physical hazards man faces on Earth today are earthquakes, tsunamis, volcanism, climatic changes leading to the growth or shrinkage of ice caps and deserts, encroachment of the sea onto the land, and extraterrestrial bombardment. In the ensuing chapters we shall look at each of these potential dangers in detail.

3 Dance of the Continents

Till a voice, as bad as Conscience, rang
interminable changes
On one everlasting whisper day and night
repeated-so;
'Something hidden, Go and find it. Go and
look behind the Ranges –
'Something lost behind the Ranges. Lost
and waiting for you. Go!'
                    Rudyard Kipling, *The Explorer*

On Sunday 31 May 1971, Señor Mateo Casaverde of the Geophysical
Institute of Peru and some friends were visiting the town of Yungay,
Peru. Yungay lies at the foot of Nevado de Huascaran, a beautiful,
21,834 foot high snow and ice capped mountain. The sky was clear
and the weather fine. As they drove past the foot of the hill on which
stands Yungay cemetery, Señor Casaverde's motor car began to
shake and rattle. He stopped the vehicle and got out to investigate. It
was only then that he realized an earthquake was in progress and,
quite naturally, he took this opportunity to make first-hand observations of the event. He saw several small buildings collapse, as well as a
bridge crossing a little river. After roughly 40 seconds the earthquake
stopped and the shaking subsided. Suddenly, his attention was drawn
to a great roar from the direction of Huascaran, where an enormous
cloud of dust had appeared. A great mass of rock and ice had broken
loose from the north side, near glacier 511, and was now cascading at
an ever-increasing speed down the narrow valley towards the villages
at the foot of the mountain. At once Señor Casaverde and his friends
began running towards Cemetery Hill, the only nearby high ground,
which was 200 yards away. They were on their way up the hill when
the wife of Señor Casaverde's friend stumbled and fell. Helping her to
her feet Señor Casaverde was horrified to see the avalanche of rock,
ice and water almost upon them. He recalls: 'The top of the wave was
curved over like an enormous breaker crashing in from the ocean and
at least 260 feet high. Hundreds of the people of Yungay were

running around in blind panic and some were trying to reach the safety of Cemetery Hill. The deafening roar was really terrifying. I reached the upper part of the hill just in time, as the debris avalanche hit the lower slopes. My friends and I must have saved ourselves by only 10 seconds.' Señor Casaverde saw a man lower down the hill carrying two small children who, as the avalanche hit him, threw the children to safety; both he, and the two women near him were swept away and never seen again.

In all 50,000 people died, 800,000 were rendered homeless and two villages were destroyed. Why? What really caused such an earth-shock? To comprehend the underlying causes of such a horrifying event it is necessary to know something of modern geological theory, and especially to understand current views on mountain building, continental drift and plate tectonics.

It is not easy for a layman to accept that the substance of all the rocks that make up the face of the Earth had a fiery beginning, any more than it is for him to appreciate that the hills and valleys which we see around us were not always there. Since our planet was formed over 4600 million years ago, the rocks upon its surface have undergone many changes; some of them have been completely remelted, some have been subjected to intense pressures and heat, while others have broken down under the relentless processes of erosion to form the components of yet other rocks. Erosion has certainly been going on since the formation of the Earth's atmosphere, and the catastrophic damage caused by sudden flood in places like India, Australia and the Mississippi valley in recent years is ample evidence that it is still going on today. It is part of a continuous recycling process which takes place at the Earth's surface and also in the Earth beneath our feet, where unseen heat sources create new molten rock, and apparently solid rocks are moulded like clay. As we saw in the last chapter, under-standing the Earth and its ancient geological processes is fraught with difficulties, for the further we go back into the geological record the more uncertain the interpretation of ancient events becomes. One of the reasons for this is that the oldest rocks are very often those which are the most altered. This means that before the Earth scientist can unravel the processes responsible for the formation of a rock, he must first decipher its true nature by identifying all the factors which brought about its alteration: the amount of addition and subtraction of different elements during periods of intense heating and pressure, the kind and degree of folding to which the rocks have been sub-jected, the amount of compression, and so on. We must ask ourselves questions like: why fossiliferous limestones, for example, which were formed in warm shallow seas, are now found several thousands of feet above sea level. Some have actually been discovered close to the summit of Everest! The evidence of the rocks is beyond question, so what are the driving forces which have heaved them from the floors of ancient oceans, thrown them into highly contorted folds and raised them up to form lofty frost-shattered peaks?

Sir Francis Bacon, English philosopher, statesman, man of letters and keen student of science, was the first person to recognize and draw attention to the remarkable geographical similarities between the outlines of the continents of Africa and South America. At the time – 1620 – the 'coincidence' aroused no more than mild scientific curiosity and doubt; it was to be over 300 years before proper scientific studies were initiated and reliable data collected. In the 1870s the voyages and scientific work of the vessels HMS *Challenger* and USS *Tuscarona* revealed the general shape of the ocean floor but, even then, the time was not ripe for the general acceptance of any ideas concerning continental drift, although the discoveries made around this time certainly laid the foundations for the scientific advances that were to come.

During the Trigonometrical Survey of India in the nineteenth century it was discovered that some measuring stations whose positions had been determined by astronomical surveying methods did not agree with those determined by triangulation. Archdeacon J. H. Pratt, a British cleric, was fascinated by the problem and suggested that the Himalayas to the north were exerting a small gravitational pull on the plumb line of the surveying instruments, thus introducing errors. After further work, Pratt concluded that the mountains, foothills and plains were 'floating' on a denser substratum and that their respective heights were dependent upon their density; since he assumed that their bases all lay at the same level, in his hypothesis it was necessary that their densities should be different. G. B. Airy, Astronomer Royal of Great Britain at that time, accepted most of Pratt's ideas, but suggested instead that the mountains, foothills and plains might each have the same density and that their relative elevation was due principally to differences in their thickness. He pointed out that a thick floating crustal block would sink deeper in the substratum, but its height above the surface would still be greater than that of a neighbouring thinner block of the same density. Today it is Airy's theory which is generally accepted. Each of the floating blocks is said to be in *isostatic* (i.e. *gravitational*) *equilibrium*.

Originally, it had been thought that both the folding of initially

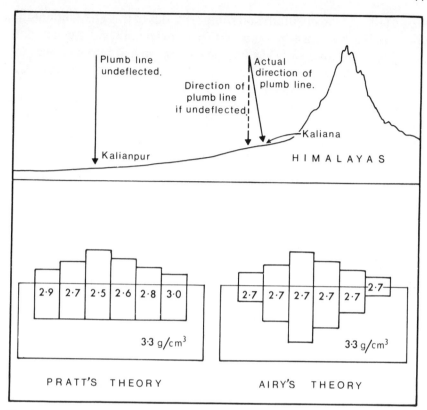

Fig. 8 The trigonometrical survey of India in the nineteenth century revealed a big discrepancy in the positions of two surveying stations, Kalianpur and Kaliana. Astronomical surveying methods gave distance values that were considered accurate, whereas triangulation techniques which depend principally upon a vertical plumb line gave different values that were at once suspect.

Pratt reasoned that the mass of the Himalayas to the north was exerting a slight gravitational pull on the plumb and thus introducing errors. He believed the base of the Crust underlying the plains, foothills and mountains was everywhere at the same level and floating on a denser substratum. In order to account for the varying heights of these different landforms it was necessary that they all had different densities, the lightest blocks forming the mountains and the densest blocks forming plains and low ground.

Airy's theory differs from Pratt's in assuming that all blocks have the same density and that the height of the hills and mountains is simply accounted for by differing thicknesses of the Crust. Airy's mountains have deep roots, whilst those of Pratt do not. Using Airy's theory it is easy to demonstrate that removal of material from the mountain block will clearly cause the block to rise to restore equilibrium.

horizontal strata and the great thickness of rocks in mountain chains was due to the Earth contracting as it cooled from an initially very hot state; but the maximum amount of shortening of the Earth's Crust which is possible by this mechanism is far less than that actually observed. In the European Alps, for example, investigation of the great folds (Plates 5 and 7) by the Swiss geologists Albert Heim and Joseph Cadisch had already, by the end of the nineteenth century, suggested that the northern and southern sides of this great mountain range had moved together and that rocks originally laid down as flat strata in a 400-mile wide sea had been compressed into a mountain range only 100 miles across. Further investigation, using geophysical instruments, has revealed that this compression created deep roots of light rock which buoy up the mountains, just as Airy had suggested. To the north-west, Scandinavia is found to be still rising after the melting of the thick Pleistocene Ice Cap around 10,000 years ago, and it is estimated that over 500 feet of further elevation is needed before gravitational equilibrium is finally reached. In order to balance such vertical movements, a horizontal movement of material is required at depth, and thus it can be concluded that a part of the Earth's crustal substratum must be capable not only of flowing beneath, but also of transporting, the overlying continental masses.

Nearly fifty years ago a German meteorologist and explorer died amidst the frozen wastes of the Greenland ice cap. In his lifetime, and for many years after his death, he was regarded as a crank because he maintained that, driven by unseen forces, the continents wandered about the face of the Earth. Alfred Wegener, the proposer of this idea, was born in Berlin in 1880, the son of an evangelical preacher. He spent much of his academic life writing about meteorology and had visited Greenland three times before he lost his life there in 1930. His ideas were so ahead of his time that they were rejected and scoffed at by the majority of scientists, yet today we know that he was right. For 25 years, the only supporters of Wegener's hypothesis were such men of genius as Arthur Holmes, the British structural geologist Sir Edward Bailey, and Wegener's devoted disciple, the South African geologist Alexander Du Toit.

When the time is ripe for a new advance in scientific theory it is not uncommon for several workers to arrive at the same conclusion simultaneously. It all starts with an idea, or rather a collection of ideas. Evidence is then gathered and sifted to determine if any of the new ideas could really work. When one does, then that idea becomes a hypothesis. The next step is to use the hypothesis to see if successful predictions can be based on it: if they can, the hypothesis is then elevated to the grand status of a theory, which is progressively refined so that it may explain how, why, when and where things happen. A time for such a new advance in scientific thinking concerning mountains, mountain building and associated earthshocks came during the 1950s and 1960s, with the development of the new science of plate tectonics. The theory of plate tectonics is remarkably simple yet, at the same time, comprehensive. It explains why earthquakes and volcanoes are found concentrated in narrow zones and why volcanoes on land usually occur within 125 miles of the sea. It also explains why some of these zones have shallow earthquakes, such as those on the San Andreas fault, and why others have deep focus earthquakes, as in the Tonga-Kermadec region. Moreover, it tells us why the ocean floor is younger than the continents which border it. What was it, then, that rehabilitated Wegener's hypothesis, brought about a revolution in the Earth Sciences and created the unifying theory of the Earth that we are presented with today?

In 1960 the late Professor Harry Hess of Princeton University, USA, presented his classic work on sea-floor spreading. Hess had been a US Navy submarine commander during the Second World War, and was particularly familiar with the Pacific Ocean. He followed the Dutch geophysicist, F. Vening Meinesz, in a belief that the sharp departure from isostatic equilibrium shown by the deep oceanic trenches close to Japan, the Philippines and the East Indies must mean that the Crust was being held down at these places by some enormous unseen force, such as the descending limb of a convection current in the Mantle. Nevertheless, before 1960, Hess was still of the opinion that the Pacific Ocean was a very old, if not primeval, feature of the Earth. For many years, Hess had been interested in the sea-mounts or *guyots* – the isolated flat-topped volcanic mounds that rise from the deep ocean floor – which are scattered all over the Pacific. He was convinced that guyots had once been normal volcanic islands truncated at sea level by wave action and interpreted the fact that their flat tops are now at varying depths below sea level as an indication of prolonged subsidence.

The 1950s saw a great expansion of oceanographic research: a world-wide burst of interest in that virtually unknown 70 per cent of our planet that lies beneath the seas and oceans. One of the first results of this work was the dredging of Cretaceous fossils from the tops of several guyots, followed by the discovery that there was far less sediment on the floor of the Pacific Ocean than had been predicted from its supposedly great age. Great submarine mountain chains, called mid-ocean ridges or rises, were being mapped out in all of the oceans of the world. It was a time for lateral thinking. Hess asked himself, 'Could the Pacific be quite young, say no older than the Mesozoic?' – and answered this question in the affirmative as a result of his knowledge of the guyots. He saw that volcanic islands truncated by wave action near the crests of oceanic rises became progressively submerged, up to several miles deep, as they were traced down the slopes of the rises. From this observation he concluded that the ocean floor must be carrying them away from the ridge crests, as if on a gigantic conveyor belt!

Hess' brilliant suggestion that the ocean floor was not static, but mobile, with continuous spreading taking place away from the ridge crests, formed the basis upon which the theory of plate tectonics was developed over the next decade. Hess was, no doubt, influenced by an

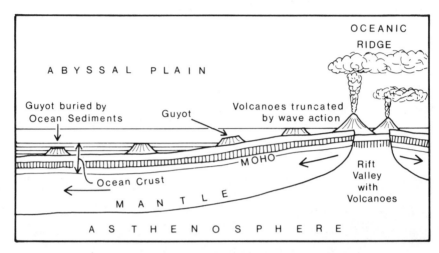

Fig. 9   Guyots are submerged truncated volcanoes. Here the volcanoes are seen forming on a median rift valley along an oceanic ridge – a constructive plate margin. Molten rock (magma) flows to the surface along cracks in the Earth's Crust. Later the magma cools and fills these cracks, thus increasing the width of the ocean floor and causing spreading of the sea floor away from the ridge. As newly created Crust migrates away from the ridge it carries the new volcanoes on its back like an enormous conveyor belt. The volcanoes are truncated by wave action to form the guyots which become totally submerged and, with increasing time, and distance from the ridge, progressively buried by ocean floor sediments.

earlier worker, David Griggs, who, while at Harvard in 1939, had suggested that fresh rock rising to the surface in the Pacific Ocean was associated with sub-oceanic mountains and that spreading of fresh rock away from these submarine peaks was contributing to the growth of the sea floor. Griggs also believed that a convection cell in the Mantle underlay the whole of the Pacific and was responsible for this phenomenon, rising in the centre of the Pacific and descending at the edge of the surrounding continents amid a flurry of earthquakes. By the 1950s it was known that these earthquakes originate on great dislocations or thrust faults which dip, as inclined planes, under the continents. Thus Hess was able to postulate that the great ocean deeps were probably due to the downwarping of the moving oceanic Crust beneath the adjacent continents.

Evidence from research on earthquake waves had already demonstrated that the Earth is made up of several concentric shells and that each shell is of different thickness and composition. From the centre of the Earth outwards, these are the Inner and Outer Core, the Lower and Upper Mantle and the Crust. In the 1950s it was found that the Upper Mantle could be divided into additional zones, one of which is a pasty, plastic zone called the Low Velocity Layer because earthquake waves travel more slowly there, due to its non-rigidity. In certain places under the oceans, rocks are heated to high temperatures in this zone of plasticity and, because hot rocks expand, they become less dense and rise relative to the cooler, more dense surrounding rocks upon which gravity has a much greater effect. When these ascending plastic rocks near the surface, part of their substance melts, leading to volcanism on the ocean floor and to intrusive activity beneath it. The new oceanic Crust that is formed in this way is temporarily lighter than the cooler rock on either side, and can be identified as an updomed elongated ridge of less dense material alongside the fissures and cracks through which its magmas ascend to the surface.

While these clues were being fitted together to form a recognizable picture, ocean floor investigations continued; geomagnetic surveys were intensified and, for the first time as a routine operation, oceanographic ships towed sensitive magnetometers to record variations in the intensity of the Earth's magnetic field automatically. The results were dramatic. They showed elongated patterns of high and low magnetic intensity, from a few miles to nearly 70 miles wide and several thousand miles long, parallel to the ocean ridges. The explanation of these *magnetic anomalies*, as they are called, was uncertain. The patterns were obviously dislocated by great submarine faults and they became indistinct towards the deep oceanic trenches and continental margins. Earth scientists everywhere were intrigued.

In 1956, a group of scientists working in England, led by Sir Edward Bullard, P. M. S. Blackett and Keith Runcorn, had found that order could be shown to exist in the apparently disordered array of the many directions of magnetism shown by old rocks, providing the assumption was made that the continents had drifted relative to each other and to the magnetic poles over long periods of geological time. In the same year, Bruce Heezen and Maurice Ewing, of the Doherty-Lamont Geological Observatory on the banks of the Hudson River, recognized that the earthquakes associated with the mid-ocean ridges were concentrated along a median rift valley in the summit of the ridge, and suggested that a mountain range 40,000 miles long encircled the world on the ocean and sea floors.

By 1962, Hess's novel explanation of the youth and origin of the ocean basins was becoming familiar to the scientific community – though it was Robert Dietz, a marine geologist at the Atlantic Oceanographic and Meteorological Laboratories in Miami, who actually first used the term 'sea-floor spreading'. In its fully developed form, the sea-floor spreading hypothesis suggested that new ocean floor material is created continuously in the crack-like rift valleys which run along the crests of the mid-ocean ridges, with fresh rock coming straight up from the Mantle to the Earth's surface to fill the ever-widening rifts. As new cracks open and are filled, so the newly created ocean floor moves progressively away from the active mid-ocean ridge.

The hypothesis was put to the test in 1963 when Frederick Vine and Drummond Matthews combined Hess' ideas of sea-floor spreading with current research on the geomagnetic reversal time scale. Matthews had been convinced as early as 1961, as a result of ocean floor dredging work, that the Mid-Atlantic Ridge was totally vol-

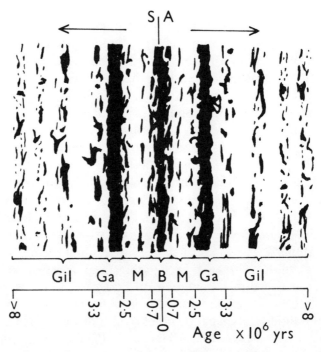

Fig. 10   Magnetic anomaly stripes: typical magnetic anomaly pattern from the Mid-Atlantic ridge south of Iceland. The parallel bands which form symmetrical mirror images either side of the ridge are produced by bands of volcanic rock becoming magnetized in the prevailing magnetic field of the Earth at the time of their cooling. Reversals in the polarity of the Earth's magnetic field are preserved and marked by reversals in the remnant magnetism of the anomaly stripes.

SA – Spreading ridge, Gil – Gilbert reversed epoch, Ga – Gauss normal epoch, M – Matuyama reversed epoch, B – Brunhes normal epoch.

canic. Vine believed that the crucial evidence in support of continental drift lay in a study of marine geology and geophysics. Working together in Sir Edward Bullard's team at Cambridge University, they pored over Hess' papers in their rooms in the old stables, comparing and matching his ideas with the new data pouring in from the oceans of the world. The results of their concentrated effort was the hypothesis that, if new ocean floor forms from volcanic rocks that are being released continuously in the deep valleys at mid-ocean ridges, it must become magnetized and imprinted with the Earth's magnetic field prevailing at the time of the cooling of these volcanic rocks below what is known as the *Curie point* (a temperature of roughly 600°C above which rocks lose their magnetism). Below the Curie point the tiny grains and rods of iron ores within cooling basalts become permanently magnetized in an orientation parallel to the Earth's field. The magnetic imprint which they receive is very faint, but with modern sensitive instruments it is easily measured and dated. It had recently become known that the Earth's magnetic field has undergone complete reversals of polarity at irregularly spaced times in the past. Thus, argued Matthews and Vine, the succession of ocean floor magnetic anomalies must be directly caused by this and, in fact, should reveal a fossil record of the reversals of the Earth's magnetic field throughout time. Indeed, they suggested that the pattern of magnetic anomaly stripes should be symmetrical and parallel on either side of the mid-ocean spreading ridges.

Many workers set out to test the Matthews and Vine hypothesis. Much of the work in this important field was carried out at the Australian National University and by the United States Geological Survey. So enthusiastic and competitive were the research teams that a whole succession of scientific papers rapidly appeared that were to give geomagnetic reversal dating a degree of sophistication hitherto unknown.

During the 1960s, after the Mohole Project to drill right through the Earth's Crust to the Mantle had been abandoned (see Chapter 2), the Joint Oceanographic Institutions for Deep Earth Sampling (JOIDES) was set up and sponsored by the National Science Foundation. The research vessel chosen was the *Glomar Challenger*, which, with its modified oil drilling apparatus, could extract rock cores nearly one mile long from the depths beneath the ocean floor. It had been predicted from the sea-floor spreading hypothesis that only young basalts would be found on the crests of the mid-ocean ridges, and that the ocean floor rocks, both volcanic and sedimentary, should show a general increase in age away from the central rifts. The

Glomar Challenger set to work in the Atlantic Ocean and very soon
proved that the hypothesis of sea-floor spreading was correct: there
was no evidence at all for ancient sediments in the central part of the
ocean basin! This was an important finding, because if ancient sediments had been found there, it would have been the end of the
sea-floor spreading and plate tectonics hypotheses.

We now know that the oceans of the world contain no sediments
older than Jurassic: more often than not there are none younger than
Cretaceous. A combination of potassium-argon radiometric dating of
volcanic rocks and palaeo-magnetic studies developed a world standard geomagnetic reversal time scale that could be applied to the
ocean floors, and, as more and better magnetic stripe data became
available, the persuasive evidence in favour of the Vine and
Matthews hypothesis steadily grew. The final piece of convincing
proof came in the form of detailed magnetic profile measurements
across the East Pacific Rise south of Easter Island. It was clear that
the magnetic stripes were symmetrical on either side of the rise, as
had been predicted. The hypothesis was correct and could now be
elevated to the grand status of a theory. Hess, Vine and Matthews had
forged the key that was to unlock the mysteries of the world's ocean
basins, so it was possible to tell how and when they had been formed
with an unprecedented degree of accuracy.

About the same time as the developments in oceanography were
being made in the early 1950s, a worldwide standardized seismographic network was being installed (see Chapter 4). Originally, it
was intended to discriminate between nuclear tests and earthquakes
in order to make a nuclear test ban feasible but, as a by-product of its
original purpose, the system also provided high quality data and
valuable new information about the focal depths and positions of
large numbers of earthquakes throughout the world. The 'quakes
were found to occur most frequently in linear belts following the great
mountain chains of the world, including the mid-ocean ridge systems,
and additionally, were located on inclined planes which continued
down from deep ocean trenches beneath the adjacent land masses or
island arcs.

J. Tuzo Wilson of the University of Toronto observed that movements of the Earth's Crust were very largely concentrated at three
types of structural features marked by earthquake and volcanic activity, namely, mountain ranges and island arcs (mobile belts),
mid-oceanic spreading ridges and major faults with large horizontal
displacements. In 1965 he suggested that these mobile belts, spreading ridges and great faults were connected in a continuous network

which divided the Earth's surface into several large rigid *plates*. He further proposed that any of these, at its apparent termination, could be transformed into one of the other two types. Great horizontal shear faults on which the displacement suddenly stops and changes direction at both ends he called *transform faults*. It was immediately apparent that transverse transform faults are of great importance along spreading ridges, causing the many step-like features and off-sets that are a prominent characteristic of those ridges.

Between 1967 and 1969 three young geophysicists, Jason Morgan of Princeton University, Dan McKenzie in Bullard's team at Cambridge and Xavier Le Pichon, a Frenchman who had spent many years at Lamont, completed the theoretical formulation of what became known as *plate tectonics*: an elegant hypothesis that could both explain and predict the motions of the crustal plates into which the present surface of the globe can be divided. Their work led to an explosion of new geological research – and a remarkable picture emerged. It was found that the constructive mid-ocean ridges, the transform faults where great crustal plates slide past each other, and the destructive ocean trench/island arc/continental margin systems where one crustal plate is overridden by another, are indeed all linked

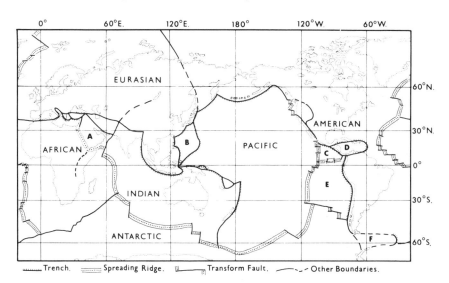

Trench.   Spreading Ridge.   Transform Fault.   Other Boundaries.

Fig. 11   Plates of the world: the Crust of the Earth is broken up into a number of plates by the combined action of oceanic ridges which cause sea-floor spreading, faults which dislocate crustal fragments over hundreds of miles, and Benioff zone/island arc trenches where excess crustal material is destroyed and recirculated. There are six major plates and six smaller ones as indicated here. The innumerable micro plates would confuse the picture and are therefore omitted from this diagram.

A – Arabian Plate, B – Philippine Plate, C – Cocos Plate, D – Caribbean Plate, E – Nazca Plate, F – Scotia Plate.

by a global network of seismicity zones which divide the Earth's surface into six major plates and about ten smaller ones. Volcanism creates new oceanic Crust as less dense rock rises to the surface at the mid-ocean ridges, or spreading axes. These ridges are sheared by further transform faults due to different spreading rates along their axes, and the actual rates of ocean-floor growth can be determined from the magnetic stripe time scales. As the ocean Crust migrates away from the mid-ocean ridge it becomes older and is eventually destroyed at the oceanic trenches in what are called subduction zones, where it plunges beneath an adjacent plate along a seismically active inclined plane. In a few years, sea-floor spreading and the unifying theory of plate tectonics were confirmed and modern geology was born.

Meanwhile, during the mid 1960s' Sir Edward Bullard and his colleagues Jim Everett and Alan Smith set about fitting the geographical outlines of the present-day continental masses together using the latest techniques and computer programmes then available at Cambridge. With certain exceptions they accomplished their task. Taking the edge of the continental shelf as the true margin of the continents rather than simply their present coastlines, South America fitted southern Africa, north-west Africa fitted eastern North America, and northern Canada fitted eastern Asia. Australia, India and Antarctica all fitted neatly together on the east side of Africa. This jigsaw construction formed a large supercontinent called Pangea, meaning 'all lands', a term originally conceived by Wegener in the 1920s – and the remarkable thing about this new 'fit' of the continents was that it was immediately acceptable to most Earth scientists. The climate of opinion had indeed changed. As Wegener had predicted, the former existence of Pangea explained away many of the anomalies that had been puzzling geologists: the problems of ancient ice ages, animal and plant evolution, migration, and many other geological and structural similarities and differences between continents. For example, it was known that mammals had evolved quite differently on different continents – the brown bear, polar bear, ant bear, elephant, kangaroo and so on – and yet they all had common ancestors way back in evolution. Those ancestors could not have swum the vast oceans which now separate the continents, and land bridges of the lengths required to allow easy migration stretched even the most pliant scientific imagination a bit too far. But, if our present-day widely separated continents were locked together as a single landmass in the past, then these evolutionary problems just do not exist. We now know that the enormous continental crustal plate of Pangea began to break up some 200 million years ago, as several spreading ridges coalesced beneath it, subdividing it into many smaller pieces with new oceans between them. Africa and South America, for example, were split apart by the growth of the South Atlantic, so the observations made by Sir Francis Bacon can be seen to have a rational explanation – continents do divide and drift across the face of the Earth! Drifting continents can join together as well as divide: Pangea is only one of many former supercontinents and it is likely that the repeated break up and reintegration of continents must play an important part in the evolution of many species of living things, not least of which is man himself.

Crustal plates on their dance around the globe may slide over *hot spots* or *mantle plumes*. Here, magma is ascending from the Mantle in a more or less discrete channel and not along a mass of aligned fissures or cracks such as pervade the mid-ocean ridges. When a hot spot occurs beneath an ocean, a line of volcanic islands, later to become sea-mounts or guyots, forms on the mobile plate as it passes over and is bulged up by the plume. Stretching away from Hawaii to the WNW and then to the north-west towards the Aleutian Trench is a long line of volcanic islands and guyots whose ages progressively increase away from the mantle plume over which the island of Hawaii is now located. During the early 1960s it had been suggested that the ocean floor in the north central Pacific had bulged up due to rising, low density mantle convection, but that ocean-floor spreading had not actually taken place. The resulting volcanic islands were then thought to have been truncated by wave action to form atolls and guyots as the entire bulge subsided after convection ceased. This static explanation – the 'Darwin Rise' hypothesis – quickly lost favour in the face of gathering and strongly persuasive evidence for more persistent mantle plumes or hot spots over which crustal plates move, carrying away long 'trains' of volcanic islands and guyots 'downstream' of the active hot spot. Mantle plumes beneath continental plates cause uplift in the form of enormous bulges or swells of the land. Rift valleys fissure the crests of these crustal swells and, eventually, extensive volcanism is localized in them. Mantle plumes underlie the Ethiopian and Kenyan rift valleys today. In the Red Sea and Gulf of Aden, new ocean floor is being produced and the continent has been split apart along what were formerly rift valleys on the crest of the Red Sea/Ethiopian continental swell. Crustal swells above mantle plumes often have a three-armed pattern of rift valleys centred on triple junctions. Whole continents may be split apart by the gradual formation of a continental spreading ridge/rift system running from mantle plume to mantle plume right across its width. An evolution of this kind was suggested for the North Atlantic Ocean by the British geologists John Dewey and Kevin Burke in 1973.

When drifting plates collide, one of several things may happen. There may be a headlong crash between an oceanic and a continental plate; such a collision is now taking place off the western coast of South America, where the eastern Pacific or Nazca plate is descending beneath the crumpled edge of the overriding South American plate and creating a great deal of friction and heat as it goes. The friction generates earthquakes which cause sudden earthshock disasters, such as that of 1971 at Yungay, Peru. Terrible as this disaster was, with its socially unacceptable death-toll of 50,000, the event was not unique. Only nine years before, in 1962, a large mass of ice, three million cubic yards in volume, fell nearly a mile from Huascaran and shattered into myriads of tiny particles. Fluidized as tiny ice crystals trapped and suspended in air, the mass travelled downslope at enormous speeds and hurled itself at eight villages. By the time the villages were reached, the mobile mass, swollen with additional debris, had increased in size to thirteen million cubic yards. Rivers were temporarily dammed and disastrous floods followed. A total of 4000 people died in this disaster. Twenty-one years earlier, in 1941, a similar tragedy partly destroyed Huaras, also in Peru, and was responsible for around 5000 deaths. Again a deadly rock and mud avalanche was to blame which, like those before and after, was directly attributable to the mountain building movements and earthquake shocks that are so commonplace in mobile belts where plates collide.

In other places, two plates may not directly collide but sidle past each other along a transform fault system, sticking and scratching as they go, and thus creating destructive earthquakes. It was the notorious San Andreas transform fault system of California that was responsible for the destruction of the city of San Francisco in 1906 (see Chapter 4).

But whether collision is direct or indirect, the people who live in these active belts of plate interaction have to accept the unpalatable fact that they are always at risk. The Cocos and Caribbean Plates are in active collision in Central America and, because of this, frequent large earthquakes and destructive volcanic eruptions are endemic to the area. Managua, the capital city of Nicaragua, the largest of the Central American Republics, is situated on the southern shore of one of the country's two largest inland lakes. The fertile volcanic soil of the region makes it an especially favourable agricultural area, and the

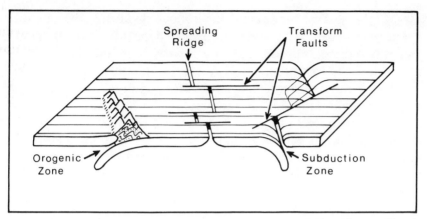

Fig. 12   Plate margins. Three kinds of plate margins are shown on this block diagram. First, the spreading ridge, or constructive plate margin, where fresh rock comes straight up from the Mantle to the Earth's surface to cause volcanism, and create islands such as the Azores, Ascension Island and Tristan da Cunha; the differences in spreading rates along this ridge cause it to break up on long transform faults. The second type of plate boundary, the subduction zone, or destructive plate margin, forms where an oceanic plate collides and underrides a continental plate; as the Nazca plate underrides the American Plate on the western seaboards of Chile and Peru. The movement of the downgoing plate against the other causes earthquakes, and generates excess heat which results in the development of a chain of volcanoes on the overhanging plate. The third type of plate margin results from the collision of two continents along an orogenic or mountain building zone. Sediments on the ocean floor are crushed between the opposing plates, folded and uplifted to form high mountain ranges like the Alps and Himalayas.

city, with a population of around 420,000, has beautiful views of the lake and surrounding volcanic peaks and enjoys an equable, balmy climate. Two days before Christmas 1972 the gala night-time street parties, dancing and general festive atmosphere in Managua were brought to an abrupt end by a series of severe earthshocks that shook, bounced and rattled the city for nearly three hours, destroying or damaging 75 per cent of the city's buildings and killing 5000 people. A further 20,000 people were injured and some 250,000 rendered homeless. For such a small and relatively poor country as Nicaragua, the economic loss was devastating. Of the 70,000 homes in the capital, over 50,000 had collapsed. Ninety-five per cent of the commercial buildings and small shops were destroyed, as were eleven large factories, four hospitals and all of the schools. Light foreshocks were felt late on the evening of 22 December. The principal earthquake (of magnitude 6.2 on the Richter Scale – see Chapter 4) occurred at half-past midnight, some two hours later. Many aftershocks followed, the two largest within an hour of the main shock, each causing substantial additional damage.

Direct collision between two continental plates results in very complex mountain building or orogenesis, and it is principally due to this process that small continents become welded together to form larger masses. Two continental plates move slowly together; between them is an ocean or a subsiding basin of deposition, which contains not only sediments eroded from the nearby continents but also lime-stones and the fossil remains of creatures that once lived in the seas bordering these converging plates. The compression as the plates move together squeezes the sediments out of the basin or sea, like putty in a vice. Heat is generated because of the intense pressures involved, and this causes some of the sediments to become altered chemically, or even to melt and travel upwards from their original position, which leads to volcanic activity. If pressure is directed from a mobile plate towards a relatively static one, then large folds may develop in the sediments on the ocean floor, overlap one another and be pushed up as a pile of young mountains onto the static plate. With continued pressure the folds become detached from their roots and may travel great distances in the growing pile as *nappes*. While this is going on, erosion is already gnawing away at the surface of the young mountain pile and sediments may be eroded off the top of an advancing nappe and deposited in front of it, only to be overridden by their own parent nappe, which itself may override, or be overridden by, yet other nappes. This is a gross oversimplification of the story, but it is due to such processes that mountains like the Alps and Himalayas owe their existence. Eventually, the two formerly sepa-rate plates become very firmly welded together along the collision line. Millions of years later, after erosion has destroyed the mountain chain, the suture will still be visible as an aligned structural belt crossing the enlarged continental plate.

The importance of mountains in elucidating the complexities of plate tectonics cannot be over-emphasized: plate collisions create mountains and therefore mountain chains should tell us both where plate collisions are occurring now and where there have been ancient plate collisions. In a global response to the developing plate tectonic regime of the Earth over the past 200 million years, the supercontinent Pangea has been dismembered and its parts dispersed. As a direct consequence mountains (orogenic belts) have grown along the western coast of North and South America, through the Aleutians and Kamchatka, down into Japan and around the Indonesian com-plex, up through Sumatra and Burma, across the Himalayas and Middle East and into the Alps. In general, the development of these mountains took place at the margins of oceans which were contract-ing in Mesozoic-Cenozoic times in response to continental drift and sea-floor spreading elsewhere.

Mountains are, of course, still in the process of formation today. In

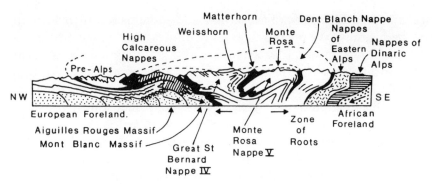

Fig. 13 A geological section through the Alps, illustrating how the mountain belt was formed as a result of the squeezing of soft younger rocks between the old rigid continental plates of Africa and Europe. As they were extruded upwards the soft rocks collapsed under gravity towards the north, producing a great pile of flat folds called *nappes*, in which great sections of strata are upside down, with older rocks resting on younger.

the Azores, new ocean floor is being generated at a constructional boundary between two great plates. Extending east from the Azores is a long active transform fault. South of Portugal the African Plate is pressing northward against the Eurasian Plate. It was the interaction of these forces in southern Portugal that resulted in the disastrous Lisbon earthquake of 1755, although there have been no further 'quakes in this area to give cause for alarm. The fault responsible was recently located with underwater cameras, and displacements of several feet were seen along it. The floor of the Mediterranean is still on the move: in response to the intense pressures generated between the northward-moving African Plate in its continuing collision with the Eurasian Plate, the western floor of this sea is going down. In the east it is rising, causing volcanism and earthquakes, one example of which was the destruction of the volcanic island of Santorini in a catastrophic eruption thirty-four centuries ago. In Cyprus, there is further evidence for the uprise of the floor of the eastern Mediterranean, for here the ocean floor has been buckled and heaved up above sea level by pressures between the plates. Some of the rocks in the Troodos Mountains of southern Cyprus are of very dense mantle material, rich in iron and magnesium, which would normally be found only at depths of eight miles or more – in the Upper Mantle deep below the ocean floor – so Cyprus is clearly the forerunner of another great Alpine mountain chain in the making, as the converging plates relentlessly crush the Mediterranean entirely out of existence. Thus, we can predict that old mountain chains formed where crustal plates collided in the past may, by analogy with the mountains that are in the process of being raised today in Cyprus and the eastern Mediterranean, include slices of uplifted oceanic rock. The presence of sub-oceanic rocks like those of the Troodos Mountains within an ancient fold belt should be a diagnostic feature, indicating that plate collision had taken place at that locality.

In certain zones of the Alps, the Himalayas and the Western Cordillera of the Americas, discontinuous slices of strongly sheared, deformed and altered basic igneous rocks have been known for some time. Harry Hess pointed to their significance as long ago as 1955. These *ophiolites*, as they are called, are now seen to be clearly analogous to the rocks of southern Cyprus, being older remnants of ocean floor caught up in fold mountains formed by plate collision. Europe and Africa are in collision along the Alps and the Mediterranean and, eventually, they will be welded together in this region by a complex of fold belts. Similarly, India is being welded to the rest of Asia along the Himalayas. In geological terms, however, these are very mobile belts.

To unravel the story of how the supercontinent Pangea was itself formed, we must examine the planed-off remains of much older mountains. One such deeply eroded old mountain belt is the *Caledonides*, a late Precambrian to Devonian zone of mountain building that can be identified in East Greenland, Norway, western Britain, Newfoundland and south-westwards through the eastern United States towards the Caribbean. Among the ancient rocks of Newfoundland, for example, there are exposures of ophiolites that can be directly compared with the ocean floor rocks of Cyprus: they too are rich in iron and magnesium and contain vast quantities of copper; they also include submarine volcanic rocks and intrusive sheets; clearly, they must have formed under very similar conditions. There are numerous exposures of ophiolites elsewhere in the Caledonides, so we can interpret this evidence as indicating that the Caledonian mountain belt represents the closure of an ancient ocean floor, followed by the final collision and welding of formerly separate continental plates.

Similar clues enable geologists to locate other ancient sutures. The Ural Mountains, which cross Russia from north to south, contain a whole series of ocean floor rock slices, just like those of Cyprus and Newfoundland. These altered ocean floor rocks are arranged in distinct linear zones along the axis of the mountain chain: they mark the line along which Siberia collided with Europe around 250 million years ago. Another ancient mountain belt that can be shown to have formed as a result of a plate collision is the *Variscan* system which extended east–west across Europe to join the Appalachians near New York before the Atlantic Ocean opened.

Taking these lines of eroded mountain belts to represent the borders of ancient continents against ancient oceans, we can begin to get some idea of what the pre-Pangean world was like. It is probable that in late Precambrian times an old linear supercontinent existed. A nucleus, consisting of South America, Africa, Antarctica, India and Australia, already in their Pangean arrangement, was joined via Arabia to North America, which itself was connected through Greenland to Eurasia. By Lower Ordovician times, around 500 million years ago, continental drift had separated Europe from North America by a small ocean. Siberia had departed from its former close proximity to Europe and separation of the Pangean nucleus from North America had been accompanied by a rotation which brought Australia, India, Arabia and the Horn of Africa closer to Siberia. Examination of the sedimentary rocks being deposited in various parts of the world at that time enables geologists to relate these continental positions to the then poles, equator and climatic belts. In North America and Europe limestones containing fossils indicative of tropical conditions suggest that these continents were near the equator while, on the other hand, a whole series of glacial deposits found in the Sahara suggest that this part of Africa must have been near the South Pole.

The small ocean that had formed between North America and Europe was short-lived. Its closure involved consumption of the ocean floor in subduction zones on either side, each of which was overlain by ocean trenches and volcanic island arcs. By Devonian times, 390 million years ago, the two continents had been reunited and firmly welded together by the Caledonian mountain chain between East Greenland, Norway, Britain and Newfoundland. Continuing drift brought Africa, South America and the rest of the Pangean nucleus towards the united North American and European continent. Collision between these two continental plates was completed by around 250 million years ago. The suture line is indicated by the Southern Appalachian Mountains of North America and the Variscan fold belt of Europe, which runs through southern Britain, France, Belgium and Germany to the Black Sea. All the time this was going on, Siberia had been moving back towards Europe and, as we have seen, collided with it to form the Ural Mountains. The supercontinent had been reformed but in a different pattern. Between the late Precambrian and the Permian the continents had performed a dance, separating and reuniting as they must have done on many previous occasions, with individual localities being carried from the equator to the poles and back again.

Northern Scotland was joined to the rest of Britain during the collision of Europe and North America that is marked by the Caledonian Mountains. Great masses of marine sediments trapped between the advancing continents were folded and uplifted to construct a belt of mountains across Scotland, Northern England, Wales and Ireland. The Ordovician volcanic rocks that are such important elements in the scenery of North Wales and the English Lake District were produced in the island arcs associated with the early episodes of plate collision. Many granites were emplaced in the fold belt during the late phases of the orogeny. Later, the Earth's Crust cracked again along a new line within this mobile belt and the plate alignments were further readjusted by lateral movements, each of which were accompanied, no doubt, by massive earthquakes. The fault line is still there; today it is occupied by Loch Ness and the Great Glen. Earthquakes still occur on this fault: in its day it must have been part of an important transform fault system like that of modern California. Lateral displacement of the Crust along the Great Glen fault totals more than 60 miles.

While the formation of Pangea was being completed, with England right in the middle of it, North America and much of Europe lay in the tropics. Life had only recently migrated out of the sea into the luxuriant swampy forests which grew along the coasts and in deltaic areas. During the growth of the Southern Appalachian and Variscan Mountains the rotted remains of the swampy forests were buried and transformed into coal in basins adjacent to the mountains. In many places the rocks containing coal seams are overlain by sand and salt deposits such as those of northern England and western Europe of Permo-Triassic times. Thus we know that extensive internal regions of the supercontinent experienced a desert environment like that of the Sahara today.

Pangea began to dismember almost as soon as it was complete: the break up started between South America and Africa during the Triassic period, just over 200 million years ago. In the northern hemisphere, North America and South America separated, then the split sent Africa away. At the same time Antarctica broke away from the African continent, carrying India and Australia with it. The first crack to develop became the site of the Mid-Atlantic Spreading Ridge, which pushed Spain and France away from North America and divided Canada from England, but it was not until around 60 million years ago that Europe and Greenland finally parted. Later still, between them, Iceland emerged on the volcanically active Mid-Atlantic Ridge. India, the fastest moving continental plate,

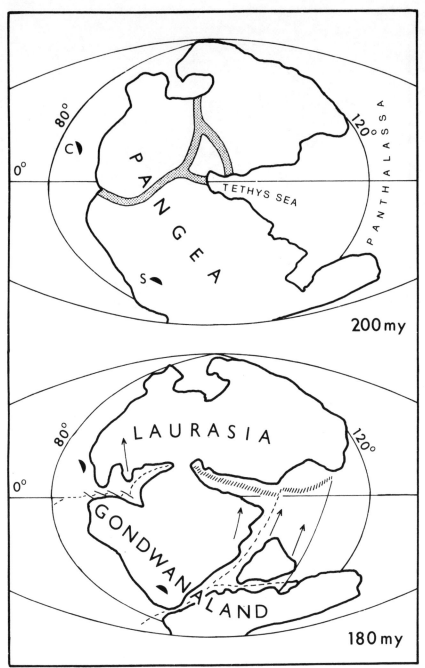

Figs 14, 15 and 16   Diagrammatic representation of the break up of the supercontinent of Pangea 200 million years ago. C – Caribbean Arc, S – Scotia Arc. Early rifting along the proto Mid-Atlantic ridge separated Pangea into the northern landmass of Laurasia and the southerly landmass of Gondwanaland 180 million years ago. By 135 million years ago the present day continental regions are recognizable, as Africa and South America separate and India detaches itself from the southerly part of Gondwanaland and begins its journey north. By 65 million

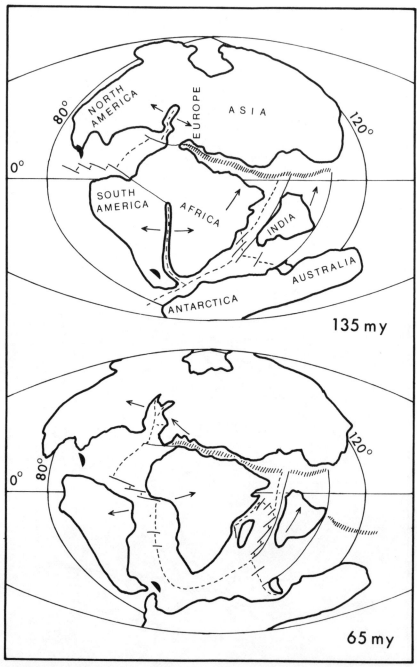

135 m y

65 m y

years ago most of the continents have separated; North America and Eurasia still remain joined while rifting steadily prises them apart. Eventually, India collides with Asia to throw up the Himalayas while similar collisions in the Mediterranean region form the Alps. Volcanism in the Caribbean region joins North and South America while the Atlantic Ocean continues to widen at the expense of the Pacific.

Assuming the continents continue to move along their present courses at their present

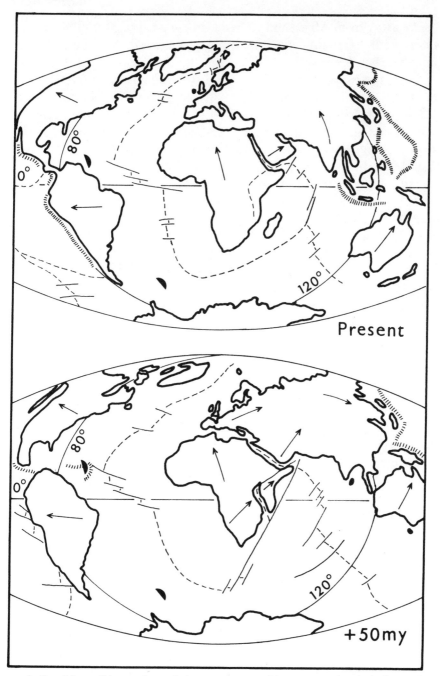

Present

+50my

speeds, then it is possible to estimate their approximate positions at some time in the future. In 50 million years' time North America and South America will have become separated, while Australia will have drifted north to collide with New Guinea. Further rifting along the African Rift Valley system will prise large segments of that continent apart, while the northward movement of Africa against southern Europe will cause Spain and Portugal to move north to collide with southern Britain and cause earthquakes and volcanism in the southern Irish Sea.

broke away from Africa and drifted north-eastwards, to collide eventually with Asia and form the Himalayas and Tibetan plateau. The mid-ocean ridge soon extended from the Atlantic, round the Cape and into the Indian Ocean, separating Australia from Antarctica. An offshoot of the mid-ocean ridge in the Indian Ocean separated Madagascar from Africa and a rift extended into the continent there, but did not get very far. Another offshoot from the mid-ocean ridge extended in between Africa and Asia, creating the Red Sea and the Dead Sea Rift, and even today the expansion of the Red Sea is slowly closing the Gulf of Oman and causing earthquakes in Iran and Turkey. It was this northward motion which threw up the folds of the Zagros mountains and was largely responsible for the formation of Iran. During the break up of Pangea, Tethys – a large oceanic gulf on the eastern side of the supercontinent – slowly began to close. Sixty-five million years ago Africa had made contact with Eurasia, leaving a thin arm of Tethys as the pre-Mediterranean and pre-Gulf of Oman. However, the expansion of the Red Sea ridge has been closing this gap during the last 60 million years.

An arm of the southern Mid-Atlantic Oceanic Ridge extended into the Pacific area 135 million years ago. Very little is known about this, but it may have been important in the opening of the East Pacific Rise, another mid-ocean spreading ridge which is slowly pushing the Nazca and Cocos Plates beneath the American Plate, causing the earthquakes and volcanoes of the Andes and Central America. A large continental plate formerly adjacent to Australia, that has been referred to as 'Lost Pacifica' may have been dismembered completely during the evolution of post-Pangean spreading ridges in the Pacific.

The last major upheaval to affect England was after the break up of Pangea, when Italy struck Europe like a battering ram and threw up the arcuate folds of the Alps. France shielded Britain during this event, but some crumpling took place in south-east England: in the Isle of Wight the strata were rotated to the vertical; the London and Hampshire regions were downfolded while the Weald of Kent and Sussex was raised up as a great dome of chalk, a tail-end of the Alps, but now unroofed and eroded down by the elements.

Finally, a great scouring of the North American and European landscape occurred during the last million years or so when the Arctic Ocean cooled and froze. Glaciers formed, due to excess precipitation of snow over melting and evaporation, and later extended across the lands of the Northern Hemisphere, carving glacial valleys and depositing glacial debris over the countryside during their retreat.

Thus we see that the ceaseless dance of the continents brings continental plates crashing into each other; sometimes they weld to form supercontinents like Pangea, but soon the enduring forces of the Earth split them apart and send them jostling on their way once more towards, no doubt, further collisions, plate destruction and the creation of further supercontinents. Man is but an inquisitive bystander who is able, during his brief lifespan on Earth, to glance momentarily at the stage where the tumultuous life of our planet is acted out. He is able, by virtue of his scientific curiosity, to unravel what has gone before and tell how the play has reached its present act. We know now that practically all violent happenings on Earth are in one way or another tied up with sea-floor spreading, continental drift and plate tectonics. The growth of sea-floor spreading ridges changes the volumes of the ocean basins; the related plate movements are responsible for further earthquakes and volcanoes; together they cause uplift and depression of the land and allow immense areas to become flooded or converted into barren desert. The movement of the continents carries them inexorably towards polar regions and into severe ice ages. In short, it both creates and destroys land and causes great changes in climate.

However academic the study of plate movement may seem, its importance cannot be minimized. It assists in the hunt for fossil fuels and in the never-ending search for other materials of economic value, and it gives us an illuminating insight into problems of earth-shock, for it controls all happenings on the Earth's Crust, other than those which may be due to extraterrestrial bombardment. Finally, it permits us to look into the future. Assuming that the plates continue to move along their present paths it is possible to give a prognosis on their future positions. After the next glacial period in the Northern Hemisphere, the widening of the Atlantic will cause more active collision between Africa and Europe, with further folding in the Mediterranean region as this sea closes. The Iberian peninsula could break away from Europe and drift north-west towards south-west Britain to cause earthquakes and volcanic eruptions in the Channel Approaches and the southern Irish Sea. The east coast of England will continue to subside, eventually drowning the London Basin and the whole of East Anglia. Even now this region is slowly but surely sinking back into the azure main. North America will probably sever from South America near the Caribbean island arc, which will probably break up. Greenland will follow Canada westwards. Canada will possibly collide with eastern Siberia while Australia will drift northwards to clash more positively with New Guinea; India may join in this collision by approaching from the west. A large fragment of East

Africa will be split off by progressive widening of the Rift Valley to form an East African Subplate, and the widening Red Sea will open into the narrowing Mediterranean. The Atlantic Ocean will expand at the expense of the Pacific Ocean, which will contract, and at some time in the future the Pacific will become the site of multiple collision and the formation of yet another supercontinent.

It is doubtful if man, as we know him, will ever see the results of this prognosis, however, for the chances of his annihilation long before then by natural or self-induced events are far too great.

4 The Unstable Earth

'Learn now the true nature of earthquakes. First you
must conceive that the earth, in its nether regions
as in its upper ones, is everywhere full of windy
caves, and bears in its bosom a multitude of meres
and gulfs and beetling, precipitous crags. You must
also picture that under the earth's back many buried
rivers with torrential force roll their waters
mingled with sunken rocks. For the plain facts demand
that earth should be of the same nature throughout.
With these things lodged and embedded in its bowels,
the earth above trembles with the shock of tumbling
masses when huge caverns down below have collapsed
through age. Whole mountains topple down, and sudden
tremors started by that violent shock ripple out
far and wide.'

> Lucretius: poem on 'The Nature of the Universe'
> addressed to Gaius Memmius in 55 BC

'The city quakes, the earth is filled with blood –'
> C. M. Verschogle, *The Deliverer*

At 5.13 on the morning of 18 April 1906 San Francisco was struck by
a severe earthquake shock. The trembling of the earth lasted for no
more than 48 seconds, yet the damage sustained by the city and
surrounding areas was so devastating that we still talk of the 'great
San Francisco earthquake of 1906'. The emotional impact, too, was
tremendous: the headmistress of a London girls' school – who was
visiting San Francisco at the time and had sheltered safely beneath a
doorway at the Opera House as masonry fell about her – offered
prayers in thanksgiving for her delivery every day at morning assem-
bly for 32 years, until she retired! The imposing City Hall, of which
the San Franciscans had been justifiably proud, was reduced to a
complete wreck (Plate 8). The colossal pillars supporting the arches
at the entrance fell into the avenue, far out across the tramcar tracks,

and the thousands of tons of bricks and debris that followed them
piled into a monstrous heap. The west wing of the building sagged and
crumpled, caving into a shapeless mass: on the north side only a steel
frame was left, gaping, bereft of its stonework cladding. All over the
city, hotels, theatres, factories and business properties were damaged
and destroyed in a similar way. The State Insane Asylum at Agnew
was demolished, the superintendent killed and many of his staff
injured. The San Francisco *Call – Chronicle – Examiner* reported on
19 April that 200 of the inmates had escaped and were roaming over
the countryside.

Many of the older, wooden framed houses and buildings that were
a feature of the city in 1906 were able to absorb the earthquake shock
more pliantly than the steel and stone structures and, as a consequence, suffered less immediate damage. However, movements on the
San Andreas fault – the principal cause of the earth tremors – broke
water mains and levelled pumping stations throughout San Francisco,
crippling the fire-fighting services. Thus, numerous small fires started
by overturned stoves, bared electric wires, broken gas mains and
fallen chimneys spread rapidly and unchecked, especially through the
wooden buildings. The Mechanics Pavilion, for example, a large
wooden building covering an entire block, survived the earthquake
and, at first, was improvised as a hospital for the comfort and care of
300 of the injured. Soon, however, as the unchecked fires spread
through the city, it had to be evacuated, and despite efforts to save it,
the big wooden structure burnt like tinder and was a flat smouldering
ruin within a quarter of an hour of the flames reaching it. The fires
raged diametrically in all directions all day and long into the following
night. By the morning of 19 April many parts of the city had been
ravaged by fire on top of the structural damage caused by the earth
tremors.

Downtown, too, everything was in ruins: not a business house still
stood. Shipping, warehouse, manufacturing and residential areas
across the city were destroyed wholesale. Confusion reigned. On
every side there was death and destruction. Women fainted and men
fought their way into buildings to rescue something from the
holocaust. Thousands were made homeless, many making their way
with blankets and scant provisions to the shelter of the Golden Gate
Park and the beach. Martial law was declared at nine o'clock and
hundreds of troops were drafted in to assist the police and fire
departments. Without mains water, the firemen used many tons of
dynamite to blast out firebreaks through whole sections of closely
spaced buildings in their attempts to check the spread of the fires.

The 1906 earthquake was centred on the San Andreas fault where it passes just west of the heart of the city of San Francisco, close to where the Golden Gate Bridge now stands. Its Richter magnitude (see page 90) is generally computed to have been around 8.3. Some 700 people died. The cost of the damage has been estimated to have been in the order of $350 to $1000 million in terms of money values of the time, mostly due to the post-earthquake fires.

Visible surface rupture of the ground occurred along the San Andreas fault for 225 miles from San Juan Bautista to Point Arena in northern California – where the fault enters the ocean – and, again, some 45 miles further north at Shelter Cove in Humbolt Country, where the vertical displacement of the upthrown side was between 7 and 8 feet. The offset of lines of trees and fences indicates that near Point Arena there was a right lateral displacement of up to 14 feet. The most extensive ground breaking in the city occurred near the waterfront and in areas built on soft Bay sediments or artificial fill. After the earthquake, it was noticed that many former straight line features crossing the San Andreas fault had become curved. The

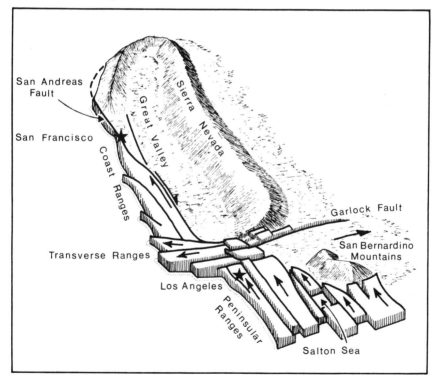

Fig. 17 Block diagram to illustrate the great complexity of the San Andreas fault system. North of the Garlock fault the complex block faulting to the south is resolved into a series of echelon faults with a clockwise movement.

American seismologist, F. H. Reid, made detailed observations of these effects and formulated what is known as the *elastic rebound theory* of earthquakes. He showed that the energy source for tectonic earthquakes (see page 85) is the potential energy stored in crustal rocks during their gradual distortion under stress. When the accumulated elastic stresses exceed the competence of the rocks, a fracture occurs, and the distorted blocks of country on each side of the fault snap back towards equilibrium. This sharp movement causes an earthquake and the energy stored in the rocks is radiated away in all directions as elastic waves travelling through the solid earth. Reid was able to demonstrate that the 1906 earthquake was caused by the faulting: it was not the earthquake that had caused the fault movements.

To the question 'Will earthquakes as powerful as the 1906 'quake shake San Francisco again in the future?' the geologist must answer with an unequivocal 'Yes!' Well might BBC-TV call their 1972 programme on earthquake risk in California 'The City that Waits To Die'. The San Andreas fault belongs to a system of great faults along which two of the Earth's major crustal plates are slowly and inexorably sliding past one another. In the brittle rocks of the Upper Crust these movements usually take the form of a series of violent jerks as rock masses fail under shearing stress. Unless we can find some way of lubricating the fault planes and so convert this jerking motion of the two sides of the fault into a smooth gradual glide past each other, then future great earthquakes in the San Francisco area are as inevitable as the passing of night and day.

The great Alaskan earthquake of Good Friday, 1964 was one of the most powerful shocks of more recent times, with a Richter magnitude of 8.6. It occurred around half-past five on the afternoon of 27 March, centred, fortunately, in a sparsely populated mountainous area north of Prince William Sound, but only 40 miles west of the southern terminal of the Trans-Alaska oil pipeline at Valdez: 114 people were killed from the direct effects of the shock, others died by drowning in the associated tsunamis (see page 80). The violence of the shaking, which lasted for four minutes, triggered numerous rockslides, snow avalanches and landslides throughout south-central Alaska and caused widespread destruction of harbour installations, railway tracks, roads, bridges and buildings of every kind. Subsidence was particularly common where the ground surface was underlain by unconsolidated sediment.

The largest city affected was Anchorage, some 100 miles from the earthquake centre. In Anchorage, parts of houses built of concrete blocks could be seen working against one another before the walls began to collapse. Whole buildings, trees, poles and masts swayed together with an east–west motion: movable objects slid back and forth across rooms as the occupants experienced a strong rolling sensation. As in San Francisco, timber frame buildings stood up to this disturbance far better than concrete constructions. In a well-to-do suburb of Anchorage, 'Turnagain Heights' situated on a high cliff overlooking Cook Inlet, clay and silt sub-soil failed as a result of the earth shaking and this led to a massive landslide carrying the development some 500 yards towards the sea. Deep chasms, into which houses, cars, trees and people tumbled, opened up in the moving ground. Elsewhere in Anchorage fractures and down-dropped blocks broke up roads and streets.

Over the whole of Alaska, as much as 80,000 square miles of land was either uplifted or dropped by several feet as a consequence of the earth movements. The docks at Cordova were elevated about 11 feet above their previous position in relation to sea level. A north-east-trending uplifted strip of the Earth's Crust over 100 miles wide and 500 miles long extended south-west of Cordova and Valdez. To the north-west a down-dropped strip of similar dimensions ran south-west from Anchorage through Homer and Kodiak Island.

Perhaps the most frightening part of the 1964 Alaskan earthquake were the great sea waves which accompanied it. The waters of Prince William Sound and the Gulf of Alaska were so disturbed by the sudden vertical displacement of the ocean floor that seismic sea waves spread right across the northern Pacific Ocean. Inland, lakes as far away as Florida were affected by *seiches*, the rhythmic sloshing of their waters due to the passage of ground waves from the earthquake. In the port of Valdez, 50 miles from the earthquake centre, sea waves devastated the entire waterfront. Within seconds of the earthquake beginning, the sea retreated and the pier broke in two, flinging warehouses and movable objects towards the retreating sea. Thirty people were on the pier waiting for a merchant ship: men, women and children staggered around the dock looking for something to hold on to. Very soon a great wave thundered back towards the shore. The 400-foot steamer *Chena* which was approaching the pier, rose like a cork some 25 feet, its bow high above the waterfront building, dropped back striking bottom, shot forward, bottomed again and was then lifted clear after heeling over as much as 50°. Two men on the ship were killed by falling cargo and another died of a heart attack. The tsunami was over 25 feet high when it surged over the waterfront, smashing all structures in its path, reducing buildings to firewood, throwing heavy trailers about and rolling up cars and lorries into twisted masses of metal. Ten minutes after the passage of the first sea wave a second surge crossed the waterfront carrying large amounts of debris. There followed a lull in which search parties looked for possible survivors in the dock area. There were none. All along the coast of North America, from Alaska to California, coastal towns were devastated by this tsunami; small ships were stranded far up the shore, the waterfronts wrecked and debris spread everywhere. As far away as Crescent City, California, 12 people were killed when the sea wave, still 20 feet high, hit that coast a good five hours later.

The greatest ever death toll from earthquake disaster is known to have occurred in China. Records of Chinese earthquakes go back to 1100 BC and are fairly complete from around 780 BC, the period of the Chou dynasty in north China. At 5 a.m. on 23 January AD 1556, near the city of Hsiăn in Shensi province, a great earthquake struck a densely populated region where the people lived mostly in caves which they had excavated in hillsides made of a soft rock called loess (an accumulation of wind blown dust). Multitudes of unfortunate sleeping peasants were buried alive when violent hammer blow earth-shocks caused these cave dwellings to collapse on top of them. Following this initial frightful catastrophe, many of the survivors died from flood, famine, disease or simply from despair: indeed, so great was the demoralization and disruption to life in the province that the official dynastic accounts of the disaster recorded 830,000 dead as the direct or indirect consequence of the earthquake.

Many other great earthquakes are known to have shocked China. On 2 September 1679, for example, an earthquake of Richter magnitude 8 occurred at a small town close to Peking. This San-ho earthquake caused damage in 121 Chinese cities and killed many thousands. In 1966 an unusual sequence of moderate earthquakes, the largest of which had a Richter magnitude of 7.2, caused widespread failures of embankments and canals and soil liquification near Shintai. Shintai is only 250 miles SSW of the area around Tientsin devastated by the great earthquakes of 27 and 28 July 1976, in which more than half a million people died. The full official accounts of the 1976 disaster have not yet been made public: sufficient is known, however, to rank it as the second greatest earthquake disaster of recorded history.

Fearsome earthquake disasters occur perennially. In the Ancient World, disasters of this kind were invariably regarded as a manifestation of the 'anger of the gods', that had to be appeased by sacrifice, communal ritual and prayers led by the local 'holy men' or priests. In Greek mythology the god of earthquakes was Poseidon, 'the Earth-Shaker', brother of Zeus and one of the most powerful of the gods. Poseidon was said to have a surly, acquisitive, quarrelsome and adulterous nature. When the three brothers Zeus, Poseidon and Hades deposed their father Chronos, they drew lots for the air, sea and murky underworld, nominally leaving the Earth common to all. Zeus won the sky, Hades the Underworld and Poseidon the Sea. With the Sea, however, Poseidon acquired mastery over the lakes and rivers, so that in a sense, the Earth belonged to him as well, since it was sustained by his waters and he could shake it at will. His claim to the overlordship of various lands around the Mediterranean led Poseidon into many quarrels with the other gods and with various mortals, during which he is described as rending mountains, causing new land and islands to appear, all often accompanied by flooding and 'tidal waves'. An earthquake god, also frequently associated with the power to cause flooding and 'tidal waves', is known in many other primitive mythologies. In Japan, for example, the earthquake god is Nai-no-Kami; in central Colombia he was known as the demon Chibchacum, who, after being defeated by the great sun god Bochica, was forced as a punishment to support the Earth on his shoulder. It is when Chibchacum tires and changes his burden to the other shoulder that there are earthquakes in the Andes and South America!

For the vast majority of us today, rooted in our early experience (i.e. in the early programming of our brain) is a belief in the 'rock-like' stability and unshaking solidity of the earth we stand upon. That the ground beneath our feet can tremble, heave and rupture in agony is unthinkable. To uninformed and uneducated peoples brought up in areas where earthquakes are unknown and even, momentarily, to trained scientific observers, involvement in a major earthquake can produce a virtually uncontrollable mindless terror and panic as the foundation of our subconscious knowledge of the world dissolves. To this terrifying subconscious reaction is immediately added a very real conscious appreciation of imminent danger as nearby buildings, cliffs and hillsides collapse, large chunks of masonry and rock crash down and the very air we breathe becomes choked with dust and noise.

Earthquakes appear to kill effortlessly; in the face of this seemingly irresistible natural violence, man feels puny and helpless. Millions are known to have died as a result of earthquakes in historical times; and still no year goes by without several great earthquake disasters being reported, harrowing pictures and accounts of the suffering and distress appearing on television and in the press, and urgent appeals for

Table 4.1   **A selection of the most disastrous earthquakes of the last one thousand years**

Year (AD)	Place	Estimated casualties
1038	Shensi, China	23,000
1057	Chihli, China	25,000
1170	Sicily	15,000
1268	Asia Minor	60,000
1290	Chihli, China	100,000
1293	Kamakura, Japan	30,000
1456	Naples, Italy	30,000
1495	Naples, Italy	60,000
1531	Lisbon, Portugal	30,000
1556	Shensi, China	830,000
1619	Dughabad, Iran	800
1667	Shemaka, Caucasia	80,000
1692	Port Royal, Jamaica	not known
1693	Catania, Sicily	60,000
1693	Naples, Italy	93,000
1707	Japan	4,900
1716	Algiers	20,000
1731	Peking, China	100,000
1737	Calcutta, India	300,000
1755	Northern Persia	40,000
1755	Lisbon, Portugal (Plate 3)	30–60,000
1759	Baalbeck, Syria	20,000
1783	Calabria, Italy	50,000
1797	Quito, Ecuador	41,000
1812	Caracas, Venezuela	20,000
1819	Kutch, India	1,800
1822	Aleppo, Asia Minor	22,000
1828	Echigo, Japan	30,000
1847	Zenkoji, Japan	34,000
1868	Peru and Ecuador	25,000
1872	Owens Valley, California	About 50
1875	Venezuela and Colombia	16,000
1883	Ischia, Italy	2,300
1886	Charleston, South Carolina	60
1891	Mino-Owari, Japan	7,300
1896	Sanriku, Japan	27,000
1897	Assam, India	1,500
1905	Kangra, India	19,000
1906	Colombia	not known

Year (AD)	Place	Estimated casualties
1906	Formosa	1,250
1906	Valparaiso, Chile	1,500
1906	San Francisco, USA (Plate 8)	700
1907	Kingston, Jamaica	1,400
1908	Messina, Sicily (Plate 10)	160,000
1915	Avezzano, Italy	30,000
1920	Kansu, China	200,000
1923	Tokyo and Yokohama, Japan (Plates 9 and 17)	143,000
1927	Tango, Japan	3,000
1930	Appenines, Italy	1,500
1932	Kansu, China	70,000
1935	Quetta, Baluchistan	60,000
1939	Concepcion, Chile	30,000
1939	Erzincan, Turkey	40,000
1940	Bucharest, Romania	1,000
1943	Tottori, Japan	1,400
1944	San Juan, Argentina	5,000
1946	Ancash, Peru	1,500
1946	Alaska and Hawaii, USA	159
1948	Ashkhabad, USSR	(?) not quoted
1948	Fuki, Japan	5,000
1948	Turkmenistan, Iran	3,000
1949	Ambato, Ecuador	6,000
1949	Tadzhikistan, USSR	greater than 10,000
1950	Assam, India	1,500
1951	San Salvador	4,000
1952	Tokaichi, Japan	600
1953	Northwestern Turkey	1,200
1953	Ionian Islands	500
1954	Orleansville, Northern Algeria	1,600
1956	Kabul, Afghanistan	2,000
1956	Santorini, Greece	none
1957	Northern Iran	2,500
1957	Western Iran	1,400
1957	Altai, Outer Mongolia	1,200
1957	Thessaly, Greece	none
1960	Southern Chile	5,700
1960	Agadir, Morocco	14,000
1960	Lar, Iran	1,000
1962	Northwestern Iran	14,000
1963	Barce, Libya	300

Year (AD)	Place	Estimated casualties
1963	Taiwan	100
1963	Skopje, Yugoslavia	1,200
1964	Niigata, Japan	26
1964	Anchorage, Alaska	114
1967	Caracas, Venezuela	266
1970	Peru	60,000
1971	San Fernando, California	64
1972	Managua, Nicaragua	10,000
1972	Jahrom, Iran	5,044
1973	Michoacan, Mexico	52
1973	Puebla, Mexico	500
1974	Lima, Peru	44
1974	Northern Pakistan	900
1975	Lianoning Province, China	several hundreds
1975	Lice, Turkey	1,900
1976	Guatemala	20,000
1976	Udine, Italy	1,000
1976	West Irian, Indonesia	433
1976	Tientsin (Tangshan), China	650,000
1976	Mindanao, Philippines	8,000
1976	Eastern Turkey	4–10,000
1977	Bucharest, Romania	1,541
1977	South Central Iran	348
1977	Sumbawa Islands, Indonesia	100
1978	Tabas, Iran	more than 15,000

financial and other aid. Yet it is not the earthquake itself that causes the appalling death toll in highly populated areas but secondary causes such as the collapse of buildings, fires, landslides and the giant waves called *tsunamis* by the Japanese. Pestilence and disease brought on by the consequent disruption of normal life and amenities after an earthquake may add considerably to the overall loss of life.

Tsunamis are often called 'tidal waves', although they have nothing to do with the tides: all known tsunamis have been created by underwater earthquakes, sea-bottom slides, land slips into water or volcanic explosions. Greek mythology records several catastrophic inundations of the land by huge waves, and it is possible that the legends of Atlantis and Noah's Flood may arise from folk memories of ancient tsunamis. The Lisbon earthquake of 1755 was accompanied by a great tsunami; Japan has suffered 15 destructive tsunamis

1   Planet Earth as seen by the Apollo-8 astronauts on 24 December 1968 during the first manned circumnavigation of the Moon. Unlike its closest neighbours in space, the Earth is a violent mobile planet with a long and complex evolutionary history. Nevertheless, life as we know it requires such a home if it is to develop and prosper.

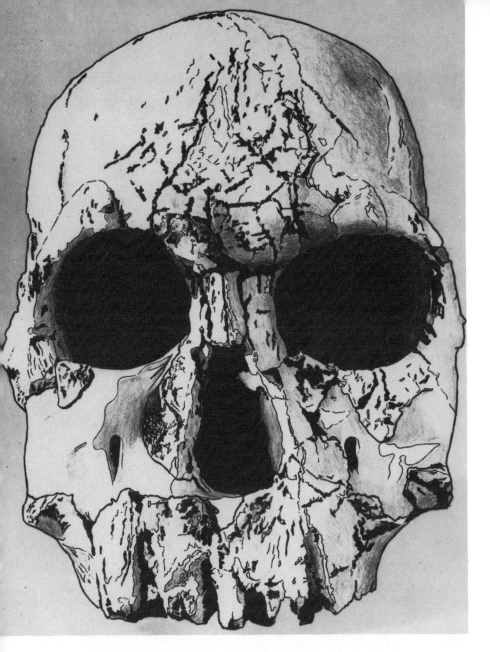

2   Our past is preserved in the fossil record. What of our future? The fossil cranium, illustrated by Malcolm Hobbs, is that of a woman who died 3 million years ago near the eastern shore of Lake Turkana in northern Kenya. East Africa at that time was probably as near to the mythical 'Garden of Eden' as we shall ever discover. Nomadic family groups of our ancestors lived a simple hunter-gatherer existence along the sparsely inhabited shores of beautiful fish-rich tropical lakes and by tree-lined rivers in a game-rich savanna wilderness. We have travelled far from this paradise: do we know where we are going?

3 An old engraving illustrates the devastation of Lisbon, the capital city of Portugal, by the great earthquake of 1 November 1755. Many of the 10–15,000 deaths that occurred on that day resulted from the collapse of churches onto their packed Holy Day congregations. The hammer-blow shocks lasted between 6 and 7 minutes. A tsunami spread right across the North Atlantic Ocean. Probably as many as were killed immediately died in the unchecked fires which burnt in the devastated city for days, as a result of riot and subsequent repression or from the diseases and pestilence that ran rife in the aftermath of the earthquake.

4 A violent thunderstorm crossed southern England during the evening of 15 August 1952. Exceptionally heavy rain fell on the upland area of Exmoor in northern Devonshire. More than 9 inches of rain was recorded in Simonsbath. From the 39 square mile catchment area of the River Lyn a flow of water nearly twice that ever recorded in the Thames at London gathered to pour through the little coastal village of Lynmouth. Overnight, the main street became the temporary channel of a raging, roaring torrent carrying all before it. Thirty-four people lost their lives in this flash flood, which destroyed houses and filled the street with huge boulders, pine trees and tons of mud. Within 48 hours the little stream was back to normal, but it took the surviving villagers many years to recover from the shock.

5   The record of the rocks: several pages of the Earth's history laid bare in a Swiss mountainside. Each thin stratum of sedimentary rock contains a record of local life and events at the time it was being deposited. The folding and uplift of these marine limestones to form part of the Alps and the subsequent erosion of deep valleys through the rock pile so formed are evidence of later events at this locality. (Schwaegalp with Saentis, Switzerland: height from top to bottom of aerial ropeway is about 2000 feet.)

6   The Grand Canyon of the Colorado River, Arizona, USA. A mile deep canyon which, with the nearby canyons, buttes and escarpments of Utah, exposes a geological section covering nearly half the Earth's history. In the inner gorge the river is today cutting into metamorphosed sediments and volcanics over 2000 million years old. These rocks were later involved in mountain building movements and intruded by igneous rocks on more than one occasion. Above an unconformity lies the tilted slab of the Wedge Series sediments, late Precambrian in age. Above them is another great unconformity cut about 600 million years ago and now buried beneath Cambrian rocks. This unconformity marks the beginning of plentiful life on Earth. The remainder of the section to the canyon rim and continued away from the rim in northern Utah, is an intermittent but very instructive rock record of the Cambrian to Recent geological history of this part of America, including evidence of numerous marine invasions of the land and many dramatic environmental changes: from desert to tropical rain forest and from freshwater lakes to volcanoes.

7   Folded rock strata near Kandersteg in the Bernese Alps, Switzerland. Mountain building deforms, metamorphoses and uplifts pre-existing rocks to produce new land areas high
above sea level: erosion begins to cut away such mountains as soon as they are uplifted.

8   Ruins of the City Hall, San Francisco, USA after the earthquake of 18 April 1906.

9   Ruins of Nihombushi, Tokyo, Japan after the devastating earthquake of 1923 in which 143,000 people perished.

10   Ruins of Messina, Italy after the great earthquake in 1908 which killed 160,000 people.

11   A dramatic drawing in the *Illustrated London News* of 1867 shows the Royal Mail steamship *La Plata* being struck by an earthquake generated tsunami at St Thomas in the Caribbean.

12   In 1973, lava relentlessly pours out of Eldjfell, the fire mountain, to destroy totally the eastern part of Vestmannaeyjar town on Heimaey island, Iceland's most important fishing port. Most of the buildings near the volcano are completely buried by volcanic ash, as was Pompeii in AD 79, 1894 years earlier.

13   Pouring down the side of Mont Pelée volcano at over 100 miles per hour the glowing dusts of destruction incinerate and kill all in their path. Such a cloud wiped out the nearby town of St Pierre, Martinique, in 1902.

since 1596, one in 1896 killing 27,000 people and sweeping away 10,000 homes; the Hawaiian Islands are struck by severely damaging tsunamis about once in 25 years, on average; and it was the enormous tsunamis generated by the eruption of Krakatoa in 1883 that caused the dreadful death toll in nearby Java and Sumatra (see Chapter 5).

The dimensions of these great waves dwarf our usual standards of measurement. Under the influence of high winds, ordinary sea waves may grow to a few hundred feet from crest to crest and travel at speeds of up to 60 miles an hour. Tsunamis, in contrast, may be as much as 600 miles from crest to crest and, in the deep ocean, they can travel at speeds of up to 500 miles an hour. They are so shallow compared to their wave length that in the open sea they can pass unnoticed. As they approach land, however, their velocity and wave length are sharply reduced while, at the same time, their amplitude may increase dramatically from a few feet to hundreds of feet. The majority have maximum amplitudes between 20 and 60 feet but a few, like the Lisbon 1755, Krakatoa 1883 and Japan 1886 tsunamis, can be well over 100 feet high as they race onto the shore. Tsunamis are more dangerous on gently sloping coasts than at places where the ocean floor has a steep off-shore slope; and they may reach mountainous proportions when constricted within a shallow V-shaped inlet or harbour mouth (Plate 11).

Generally, the first wave is a rather sharp swell that is not too alarming. This is followed by a tremendous suck of water away from the shore as the first great trough of the tsunami wave train arrives. Then comes the first of the huge waves, sweeping all before it as it crashes onto the land. The rather long interval – 15 minutes to one hour – between waves often lulls people into a false belief that the tsunami has passed, and many additional deaths have resulted from people returning to their inundated homes too soon, before the last of the killer waves have arrived.

Severe earthquakes, which would be disastrous or even catastrophic in the populated areas of the earth, are really quite frequent, occurring perhaps as often as one every two or three weeks on average but, as the majority of these 'quakes originate in the oceans, beneath the continental slopes and submarine mid-oceanic ridges, or in remote mountain districts on land, they cause little damage directly related to human life and thus pass virtually unnoticed and unreported. Nevertheless, earthquakes are the most persistent form of earthshock to which the human race is subjected, and a brief review of the more notable seismic events that have shaken the earth in *any* one twelve-month period reveals the full horror of this natural killer.

Table 4.2   World earthquake record from June 1976 to May 1977

Date	Richter magnitude	Location	Death toll
1976			
3 June	6.8	Solomon Sea, South Pacific Ocean	
7 June	6.5	Philippines	
7 June	6.6	Offshore Mexico, Pacific Ocean	
20 June	7.2	Indonesia	
25 June	7.1	West Irian, Indonesia	37 villages were damaged, massive earthslides buried an unknown number of people, 433 dead were recovered.
26 June	6.6	Talaud Islands, Indonesia	
11 July (16.45)	7.0	Panama-Colombia border	
11 July (20.41)	7.1	Panama-Colombia border	
17 July	6.6	New Britain, Papua New Guinea	
27 July	8.2	Tientsin, China	The greatest earthquake since the Alaskan earthshock of 27 March 1964 (magnitude 8.6). Tremendous damage and a loss of life of the order of 650,000 persons. An exceptionally strong after-shock occurred 15 hours later (listed below) and added to the damage already sustained in this densely populated industrial and mining region of China.
28 July	7.9	Tientsin, China	
2 Aug.	6.9	New Hebrides, south-west Pacific Ocean	
16 Aug.	6.9	Kansu-Szechwan, China	
16 Aug.	8.0	Celebes Sea, south of Mindanao, Philippines	This submarine earthquake generated a tsunami with wave heights of up to 50 feet. More than 8000 people died when the tsunami washed over the coastal areas of south Mindanao and nearby islands.
17 Aug.	6.7	South coast of Mindanao, Philippines	
21 Aug.	6.6	Kansu-Szechwan, China	
15 Sept. (03.15)	6.2	Udine, north-east Italy	
15 Sept. (09.21)	6.0	Udine, north-east Italy	The two Udine earthquakes and as many as 40 smaller shocks occurred in an area just recovering from a 'quake which killed 1000 people in May 1976 and, although no further deaths were reported, reconstruction work suffered a serious setback.

Date	Richter magnitude	Location	Death toll
30 Sept.	6.5	Kermadec Islands, South Pacific Ocean	
29 Oct.	7.1	Jayawijaya Mountains, West Irian, Indonesia	At least 133 persons were killed by this earthquake. Many were refugees from a previous earthquake (25 June 1976).
7 Nov.	6.3	North-east Iran	16 people were killed and 32 persons injured.
7 Nov.	6.8	Mindanao, Philippines	
15 Nov.	6.5	120 miles east of Peking, China	Described as a further aftershock of the destructive 'quake of 27 July 1976. No further casualties reported.
18 Nov.	6.5	Solomon Islands	
24 Nov.	7.4	Eastern Turkey	Between 4000 and 10,000 people were killed and 30,000 rendered homeless by this great earthquake in Asia Minor.
30 Nov.	7.3	Northern Chile	
20 Dec.	6.5	West of Vancouver Island, British Columbia, Canada	
1977			
6 Jan.	6.6	Bismarck Sea, north-east Greenland	
31 Jan.	6.1	Tadzhik SSR, USSR	Considerable injuries and damage in the Isfara area.
4 March	7.2	Eastern Romania	1541 people were killed, 11,275 injured and there was extensive damage to property, mainly in the capital, Bucharest.
18 March	6.8	Luzon, Philippines	
21 March	7.0	South-east Iran	More than 100 people were killed.
2 April	7.5	Tonga Islands	
6 April	6.0	South-central Iran	In an area greater than 75 square miles there was severe damage to property, 348 people were killed and over 200 badly injured.
20 April (23.13)	6.7	Guadalcanal	
20 April (23.43)	7.4	Guadalcanal	
21 April (04.24)	7.6	Guadalcanal	
12 May	5.8	near Peking, China	

For every major earthquake listed above there were 20 to 50 minor tremors that could be felt by people and many more so slight that they were only recorded on instruments. Specifically volcanic tremors are not listed. A concentration of severe tectonic earthquakes in certain earthquake-prone regions is suggested, as is the common occurrence of further 'quakes in areas already devastated by a severe earthquake.

An earthquake is simply the passage of vibrations through the solid
Earth; their transmission inside the globe and across its surface can be
compared to the spreading pattern of waves on the surface of a quiet
pool of water when a stone is thrown into its centre. Earthquakes can
be of every intensity from the faintest tremblings detected only by
very sensitive instruments to the passage of giant waves of the land
surface some 20 to 30 feet from crest to crest and with an amplitude of
up to a foot or more. Vibrations in solid bodies are set up by sudden
blows or tears, or by the scraping together of two rough surfaces. The
vibrations of musical instruments, produced, for instance, by ham-
mers tapping wires or membrane, as with pianos or drums, or by the
scraping of rough surfaces together, as in the playing of a violin, are
transferred to the surrounding air and some become audible, while
others are of too high a frequency to be heard; in a similar way
earthquakes may be accompanied by both audible and inaudible
noise. It is this 'unheard' noise, caused by atmospheric vibrations
outside the range of audibility of the human ear, that often disturbs
dogs, fish and other animals immediately before major earthquakes.

Earthquakes originate where the rocks of the Earth's Crust and
Upper Mantle rupture under stress; where rocks move one against

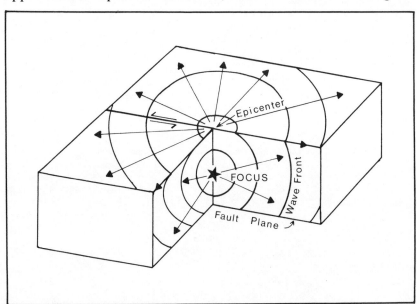

Fig. 18   Anatomy of earthquakes: crustal movement causes immense stresses to build up in
rocks. When the stress reaches a particular value the rock snaps along a fault and releases the
accumulated energy as an earthquake. These earthquake vibrations radiate out from the
fracture point or focus. The point on the surface directly above the focus is known as the
epicentre. From here another kind of wave radiates out parallel to the surface. It is this wave
that is responsible for the damage which is caused by earthquakes.

the other along great fracture planes (faults), cutting the outer layers of the globe; where molten volcanic rock is moving upwards towards the surface; in the gas discharge at volcanic throats; and, in fact, wherever there is an explosion of any kind, whether volcanic or where an atom bomb or other explosive charge has been detonated. Naturally, the intensity of the earthquake vibrations decreases away from their source.

The majority of natural earthquakes, including the most disastrous and all catastrophic examples, can be distinguished as *tectonic* earthquakes and result from the rupture of rocks under stress, possibly accompanied by subsequent movement on faults. There are two main kinds of tectonic earthquake: normal or shallow earthquakes, which originate in the Earth's thin surface Crust; and deep-focus earthquakes which occur at depths of up to 500 miles in the Earth's Mantle. The rocks of the Mantle have a noticeably greater density than those of the Crust and, because of this, the velocity of transmission of earthquake waves is considerably greater in the Mantle than in the Crust.

Volcanic explosions and violent subvolcanic movements, on the other hand, result in another, more frequent, but rarely exceptionally intense, group of earthquakes which usually originate between a depth of about 50 miles and the surface. The movement of molten rock through the Upper Crust immediately prior to and during some kinds of volcanic eruption produces low pitch rapid vibrations of the Earth known as *harmonic tremors*. These tremors and many other weak vibrations in the earth, such as those caused by the passage of trains and heavy lorries, avalanches, landslides and the collapse of caverns or mines, can be registered and recorded on seismographs but are not 'earthquakes' in the popular meaning of the word.

The intensity of an earthquake can be measured in a subjective manner on various scales such as the *Rossi-Forel* and the *Mercalli Intensity Scales*. Not only will the intensity of any one earthquake decrease with distance from the source; its effect at any one locality will depend also on the nature of the underlying bedrock, while its destructive power in human terms will be related to the population density and the type of buildings erected in the area affected by the earthquake. The most used of the intensity scales, the modified Mercalli Scale, measures the intensity of an earthquake in a number of degrees from I, at which strength it will only be detected on a seismograph, to XII, a catastrophic earthquake causing total destruction. Although these earthquake intensity scales may appear unnecessarily detailed and perhaps unconsciously amusing at first glance, they are, in fact, a graphic record of the actual effect of an earthquake as felt by people living in an area in which a 'quake is detectable.

Table 4.3 Subjective scales of earthquake intensity

(A) *Rossi-Forel Scale*

I	*Microseismic shock*. Recorded by a single seismograph or by seismographs of the same model, but not by several seismographs of different kinds; the shock felt by an experienced observer.
II	*Extremely feeble shock*. Recorded by several seismographs of different kinds; felt by a small number of persons at rest.
III	*Very feeble shock*. Felt by several persons at rest; strong enough for the direction or duration to be appreciated.
IV	*Feeble shock*. Felt by persons in motion; disturbance of movable objects, doors, windows, cracking of ceilings.
V	*Shock of moderate intensity*. Felt generally by everyone; disturbance of furniture, beds, etc., ringing of some bells.
VI	*Fairly strong shock*. General wakening of those asleep; general ringing of bells; oscillation of chandeliers; stopping of clocks; visible agitation of trees and shrubs; some startled persons leaving their dwellings.
VII	*Strong shock*. Overthrow of movable objects; fall of plaster; ringing of church bells; general panic, without damage to buildings.
VIII	*Very strong shock*. Fall of chimneys; cracks in the walls of buildings.
IX	*Extremely strong shock*. Partial or total destruction of some buildings.
X	*Shock of extreme intensity*. Great disaster; ruins; disturbance of the strata, fissures in the ground, rock falls from mountains.

(B) *Mercalli Scale* (as modified to suit modern western life styles and conditions)

O	Only registered on seismographs.
I	Not felt, except by a few persons in very favourable circumstances, such as those lying down on hard ground.
II	Felt by persons sitting down or in bed, particularly in high buildings. Delicately suspended objects may swing.
III	Felt generally (but not by everyone) indoors, again especially on the upper floors of buildings. All hanging objects begin to swing. The vibration is like the passing of vans or light lorries and quite often it is not recognized as an earthquake.
IV	Very noticeable indoors; hanging objects swing strongly and a vibration is felt that is comparable to the passing of heavy lorries or trucks or, alternatively, there may be the sensation of a jolt like that of a

heavy iron demolition ball striking the walls of the house. Standing motor cars rock, glasses clink, crockery rattles, doors bang and walls make creaking sounds.

V   Felt by most people outdoors as well as indoors. Light sleepers awakened. Liquids are disturbed and some are spilled. Small unstable objects fall over or are displaced. Doors swing open and closed. Pendulum clocks stop and start or change their rate of swing. Disturbances of trees, poles and other tall objects may be noticed. A few windows and some crockery may become broken. Vibration, jolts and noise generally recognized as being due to an earthquake.

VI   Felt by everyone, whatever they are doing. Many people are frightened and run outdoors. Persons appear to be walking unsteadily. Many windows, dishes and glassware are broken. Furniture is moved and light furniture is overturned. Pictures fall off walls and books off shelves. Smaller church bells ring and trees are visibly shaken. Weak plaster is cracked and some tiles fall.

VII The shock is noticed by everyone, even the drivers of fast moving motor cars, and recognized as a severe earthquake. Many people find it difficult to stand; as soon as they can everyone runs outdoors. Hanging objects quiver violently. Furniture is broken. There are great falls of glass, plaster, loose bricks, stones, tiles and cornices; walls crack and crumble. Damage is slight in well-designed and constructed buildings, only moderate in ordinary houses but may be severe in poorly built homes and in badly sited and designed structures generally. Small slides and caving-in occur along sand and gravel banks; even concrete irrigation ditches may be damaged. Larger church bells ring.

VIII People begin to panic. The steering of moving motor cars is affected. All man-made structures are damaged, but in those specially designed and constructed to withstand earthquake risk, the damage will be slight or negligible. In less well-designed and constructed buildings, masonry will be badly damaged, with the twisting and fall of chimneys, factory smoke stacks, columns, monuments, towers and walls. All window glass is broken and falls to the ground below. Frame houses move on their foundations and their panel walls may be thrown out. Heavy furniture is overturned. Branches are broken from trees. Cracks appear in wet ground or on steep slopes and sand and mud are ejected in small amounts. The flow or temperature of springs and wells changes.

IX   General panic ensues. All structures and masonry, except heavily reinforced concrete, will be severely damaged, with partial or complete collapse. General damage to the foundations of all buildings. Serious damage to reservoirs and dams. Underground pipes are broken. Conspicuous cracks appear in the ground.

X   Most masonry, frame structures and buildings of all kinds are completely destroyed along with their foundations. Only a very few well-constructed wooden buildings survive. Dams, dykes and embankments are breached. Large landslides occur. Water is thrown (slopped) onto the banks of canals, rivers and lakes. Sand and mud are shifted horizontally on beaches and flat land. Railway lines are slightly bent.

XI Railway lines are greatly bent, often in sinuous forms. All underground pipelines become unserviceable. Very few, if any, buildings remain. Great landslides and floods sweep into valleys.

XII Catastrophic disaster: the damage is virtually total. Towns and cities are razed to the ground. Large rock masses are displaced. Lines of sight and level are altered. Objects are thrown into the air. The ground is seen to rise and fall in great waves. Uncontrollable panic and despair affect the populace.

(C) *Mercalli Scale* (as modified to suit the social and environmental conditions of Japan)

O Not felt by humans, only registered on seismographs.

I Slight; felt only feebly by some persons at rest, those specially sensitive to earthquake.

II Weak; felt by most persons, causing light shaking of windows and Japanese latticed doors (Shoji).

III Rather strong; shaking of houses and buildings, heavy rattling of windows and Japanese latticed sliding doors, swinging of hanging objects, sometimes stopping pendulum clocks and moving liquids in vessels. Some people are so frightened as to run out of doors.

IV Strong; resulting in strong shaking of houses and buildings, overturning of unstable objects, spilling of liquids out of vessels.

V Very strong; causing cracks in brick and plaster walls, overturning stone lanterns and gravestones, etc.; and damaging chimneys and mud and plaster warehouses. Landslides observed in steep mountain areas.

VI Disastrous; causing demolition of more than 1 per cent of Japanese wooden houses; landslides, fissures on flat ground which are occasionally accompanied by the spouting of mud, sand and water in low fields.

VII Ruinous; causing the demolition of almost all houses; large fissures and faults are observed cutting the ground.

Initially, earthquake studies were restricted to the gathering of eyewitness accounts and data on surface damage, the displacement of objects, orientation of cracks, the occurrence of earth movements, seiches and tsunamis. By comparing the apparent intensity experienced at various localities, maps could then be prepared showing the distribution of isoseismal areas (that is, areas in which the earthquake had the same intensity). From the pattern of isoseismal contours on such a map it is usually possible to define the small central area or *epicentre* in which the earthquake was at its most intense. Careful study of the way objects had fallen or were displaced, and of the orientation of cracks in buildings and in the ground were used to confirm or supplement these conclusions. It can then be assumed that the actual focus from which the shock waves originate is immediately below the epicentre but at an unknown depth. As we have seen, tectonic earthquakes are caused by the explosive rupture of strained rocks under stress or by further sudden rasping movements along fractures (faults) present as the result of failure of the rocks: thus some epicentres may map as oval or elongate areas aligned along the underground fault upon which the *focus* or line of foci are situated. Within such an elongate epicentre the greatest damage may not occur immediately above the focus, but may be related to movement further along the fault in response to the initial explosive failure.

In the time of the Han Emperors (202 BC to AD 220) an instrument for the detection of earthquakes was in use in China. It consists of a bell-shaped object with eight dragons, each of which looks out in a different direction with a ball in its jaws. An earthquake of a certain strength would make the ball drop from the jaws of the dragon facing the direction from which the shock came. In Europe a variety of similar instruments were devised in recent centuries: they ranged from many spouted bowls filled with mercury to elaborate wooden 'Christmas tree'-like objects with delicately balanced cursors spread along their many branches. Instruments of this kind are called *seismoscopes*: they can be used to detect the occurrence of earthquakes and to give a crude estimate of the direction and relative intensity of the source of the disturbance. In 1853, Luigi Palmieri, Director of the Vesuvius Volcanological Observatory, made an instrument that, in addition, could note the exact time of arrival of a shock and measure its duration. In 1880, John Milne, an Englishman working in Japan, improved on this idea and produced the first modern instrument capable of measuring and timing all of the ground vibrations associated with an earthquake. Today, a network of such instruments, called seismographs or seismometers, are located around the world. With the information derived from this network of seismographs, scientists can locate both the epicentre and the depth of focus of any natural earthquake, small or large, with great accuracy. Man-made explosions and other events are also recorded on seismographs. In the last two decades some very sophisticated arrays of instruments have been designed and constructed that are capable of detecting and locating very weak earth vibrations from distant sources, including even those originating from underground nuclear tests taking place on the other side of the globe.

While the Mercalli and other similar scales give a detailed, if subjective, account of the intensity of an earthquake at any one point, it is now more common to measure the actual 'size' of an earthquake on the *Richter Scale*. This relates the measured amplitude of an earthquake, as recorded on seismographs, to the total energy expended. It is an open-ended scale and progresses logarithmically; roughly speaking, an increase of one unit on the Richter Scale means that the earthquake is 30–35 times as great as at the number below. As expressed on this scale an earthquake of magnitude 4.5–5 will only cause local damage, while a magnitude of 7 or over signifies a major earthquake. Possibly the greatest earthquake known, the Lisbon earthquake of 1 November 1755, had a magnitude of about 8.9 on the Richter Scale; only the very largest earthquakes exceed 8.4.

The Crust of the Earth is almost everywhere under some sort of stress. The greatest accumulation of stress occurs, as we saw in Chapter 3, along the boundaries between crustal plates, but even within plates, stresses may accumulate to beyond the elastic competence of the rocks in many different situations. Beneath volcanoes, it is the forceful upward movement of hot magma and the explosive release of volcanic gas that cause rock failure. Thus it is that, at any one time, in many parts of the world, there will be large or small areas of the Crust in which the elastic stresses have accumulated to very near the point of rock failure and rupture. In these circumstances, the advent of quite small additional stress will be sufficient to trigger off an earthquake. This is why the necessary change in the crustal stress pattern in the aftermath of a large earthquake can start a chain of larger and/or smaller shocks elsewhere, and why the very small alterations to crustal stress values engendered by the passage of deep cyclonic depressions or by the tides, especially by the tides of the solid earth, can be sufficient to trigger whole sequences of earthquakes. The tides of the solid earth are small sequential variations in the diameter and shape of the globe caused by the combined gravitational pull of the Sun, Moon and other planets on the body of the Earth. Long-term variations in the amplitude and periodicity of these Earth tides are related to the relative motions of the planets. At times when they are enhanced as a result of a suitable alignment of the planets and the Sun, there may be a noticeable increase in earthquake activity: the influence of the giant planet Jupiter may be particularly instrumental in this matter: the so-called 'Jupiter effect'.

Earthquake vibrations travel through the solid Earth in two forms: a compression or push-pull vibration in the direction of propagation

(P-waves) and a transverse, distortional or sideways oscillation (S-waves). P-waves travel faster than S-waves and arrive long before the S-waves from distant earthquakes. Both types of vibration are either refracted or reflected at the density boundaries that occur in the rocky layered structure of the Earth. This fact has enabled us to use both artificially induced and natural vibrations to probe the internal structure of the Earth. Artificial shocks are widely used in the study of the rock structures in oil fields, coalfields and to help find new underground ore bodies. Studies of natural earthquake wave propagation have enabled us to delimit and define the major layers of which the Earth is composed. The fact that S-waves are not transmitted through the Core of the Earth confirms us in our belief that, although under tremendous pressure, the Core consists of matter in a hot fluid state.

International co-operation over many years has resulted in a great volume of data gathered by the world's network of seismographs being readily available for study. Thus many facts about earthquakes and their distribution can be derived and, based on them, some startling conclusions have been reached regarding the nature, dynamic mechanism and history of the outer crustal layers of the Earth. Earthquakes do not occur at random around the globe. There are a small number of very clearly defined earthquake 'zones' within which the vast majority of all natural 'quakes both small and large occur. As we saw in Chapter 3, there is a direct connection between mountain chains, earthquake belts and the location of many of the most violent of the world's active volcanoes. It is not a coincidence that the Circum-Pacific earthquake belt very largely coincides with the Circum-Pacific volcanic 'ring of fire' and with the belt of young fold mountains that encloses the Pacific Ocean. Other major earthquake belts include the Alpine-Himalayan belt that runs east from the Mediterranean through Asia to join the Circum-Pacific belt in the East Indies; the Pamir-Baikal zone of central Asia; a wide triangular active area in eastern Asia, enclosed by the Pamir-Baikal zone, the Alpine-Himalayan and Circum-Pacific belts; the rift valley zone of Africa; the Hawaiian Islands and the great oceanic rises, including the mid-Atlantic and mid-Indian Ocean rises. The major earthquake belts also coincide with crustal plate boundaries (see Figs 12 and 19). At the ocean-floor spreading centres, at Hawaii, beneath active volcanic areas and within areas such as the Kenya Rift, many shallow earthquakes are directly connected with volcanism, both with upward movements of magma and with actual explosive eruption (this type of earthquake will not be considered further in this chapter). Shallow tectonic earthquakes originating from rock failure and fault move-

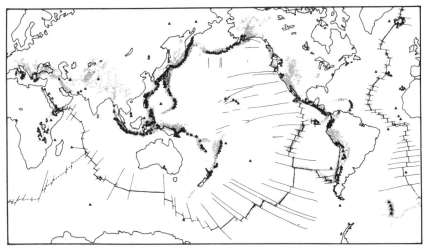

Fig. 19   Volcanoes and earthquakes of the world: the relationship of volcanoes and earthquake zones is clearly seen to correspond to major plate boundaries of the world (see Fig. 11). It
is here that the interaction of growing or colliding plates produces destructive earthquakes and
catastrophic volcanic eruptions. The chain of volcanoes on the Pacific ring of fire coincides with
destructive plate margins (Benioff zones).

ments occur everywhere, but are most frequent along active plate
boundaries, in rift zones, along transform faults and in old block
mountain areas where rejuvenation and/or readjustment are taking
place. Deep-focus tectonic earthquakes only occur along plate boundaries where subduction is taking place. For many years it was realized that inclined planar zones of deep-focus earthquakes, going
down from 50 to 500 miles in depth, were present beneath certain
island areas and mountain ranges – they were called *Benioff zones*.
With the development of the plate tectonic theory it became clear
that Benioff earthquake zones represent the tectonic track of an
overridden crustal plate as it plunges down into the Upper Mantle.
Good examples of simple Benioff zones underlie the Japanese Arc
and the Aleutian Islands. More complex Benioff zones must underlie
all young mountain chains.

   Nowhere in the world can be said to be totally earthquake-free.
The modern inhabitants of the ancient English cathedral city of
Canterbury in Kent, for example, probably believe that they live in a
part of the world that never has earthquakes. Yet, while it is true that
Kent has been free from strong earthquakes for many years, it is
certainly not true that earthquakes are unknown in that area. On 1
June 1246 'there happened so great an earthquake in England that
the like had seldom been seen or heard. In Kent, it was more violent
than in any other part of the Kingdom, where it overturned several
churches.' On 21 May 1382 another quite strong earthquake struck

the Canterbury area, 'fearing the hearts of many, sinking and throwing down many churches'; on 6 April 1580 an earthquake that was also felt in France and Belgium shook much of England. In East Kent, three distinct shocks were felt, at 6 p.m., 9 p.m. and 11 p.m. At Dover, part of the 'white cliffs' fell into the sea, carrying away a portion of the castle wall. Falls of masonry also occurred at Saltwood Castle and Sutton Church, in Kent. In the same year, just after midnight on 1 May, 'an earthquake was felt in divers places in Kent, namely at Ashford, Great Chart, etc., which made the people rise out of their beds and run to their churches, where they called upon God by earnest prayer to be merciful unto them.' On 1 June 1756, a shock at Ashford was accompanied by a noise like the report of a cannon; a shock lasting eight seconds and attended by a distant rumbling noise was felt throughout East Kent on 27 November 1776; a strong shock that occurred on 2 March 1831 was felt around the coast from Dover to Margate and a ground trembling accompanied by subterranean noise lasting four or five seconds was reported from Sandwich in Kent on 19 October 1848. Without doubt, further strong earthquakes will occur in Kent sooner or later; and when they do, unprepared and unenlightened people experiencing such an earthshock for the first time will be as frightened as their predecessors were in 1382 or 1580.

The inhabitants of other parts of the British Isles have not escaped earthquake shock. London has been shaken many times: chimney stacks were thrown down and masonry fell from St Paul's, Temple and Christ Churches during the widely felt earthquake of 6 April 1580; a London apprentice was killed by the stones falling from Christ Church. In 1750, there were six earthquakes in London between February and June: in the worst of these, on 19 March, many chimneys and some houses collapsed as a result of the 'strong and jarring motion attended by a rumbling noise, like that of distant thunder'. Stones fell from the new spire of Westminster Abbey, dogs howled, sheep and other animals were agitated and fish leapt out of ponds in which the water slopped to and fro. The greatest earthquake known in Britain in recent years was centred near Colchester in Essex and occurred at 9.18 a.m. on 22 April 1884. In Colchester more than 400 buildings (including four churches and six chapels) were damaged. In the nearby village of Peldon, every house and cottage suffered damage. Chimney stacks were thrown down and, altogether, within an area of 150 square miles more than 1200 buildings had to be repaired. According to observers who experienced the Colchester earthquake the shock consisted of two distinct parts separated by two or three seconds.

It is likely that all of the natural earthquakes that have occurred in Britain during the last 1000 years have been caused by movements on faults. Not surprisingly, then, many of these earthquakes are known to have been centred on such great dislocations in the Earth's Crust below Britain as the Great Glen Fault, which bisects the North-west Highlands of Scotland from north-east to south-west; the Highland Boundary Fault which runs from Stonehaven to Rothesay; and the Dinorwic Fault in the Menai Straits which separates the ancient island of Anglesey from the rest of North Wales. Where the old basement rocks of Britain are deeply buried beneath the younger sedimentary cover, as in much of south-east England, it is not easy to identify the individual faults that may be responsible for an earthquake. We know, however, that most of the earthquakes that have shaken Kent must result from small stress accommodation movements on one or more of the multitude of faults that cut complexly folded Carboniferous and older rocks that are present below a relatively thin up-arched cover of younger strata. The complexities of the underground rocks of Kent are well known because they include a buried coalfield that has been extensively explored by mining and bore-holes. This belt of fractured and folded rocks beneath Kent is part of the denuded root zone of an ancient mountain belt: all that is left of a chain of great mountains that dominated southern Britain and nearby Europe for nearly 100 million years between 300 and 200 million years ago before they were planed off by erosion. Gradual distortion of the European continent crustal plate as it is remorselessly driven across the face of the globe by the forces generated at ocean-floor spreading axes causes various stresses to build up within its brittle near surface layer. At any locality, when these stresses reach a critical level, rocks break and faulting, movement and readjustment must occur. This may take the form of a renewal of movements on old fault patterns or the creation of new faults. In either case the result is an earthquake or series of related earthquakes. All modern British earthquakes are attributable to this cause.

This does not mean that distant earthquakes are not felt in Britain. In fact, the effects of the great Lisbon earthquake of 1 November 1755 were noticeable as far as Scotland. At around 9.30 a.m. Loch Lomond suddenly and without the least gust of wind rose 2 feet 4 inches against its banks then just as suddenly withdrew to a similar distance below its current level. A similar oscillation in level continued every five minutes for three-quarters of an hour, then began to wane. By 11 a.m. the loch was still again. Seiches of this kind were seen in canals, ponds and lakes all over Britain. The seismic sea waves did not reach Britain until the afternoon. In Penzance Bay the height of the greatest wave was 8 feet when it arrived at around 3 p.m.

The destruction of Lisbon in AD 1755 (Plate 3) was a disaster which had a profound effect on European thought in the latter part of the eighteenth century. Voltaire was jolted out of his former optimism and complacency and his adherence to the view that 'all is for the best in this the best of all possible worlds' to become the influential sceptic of his later years, who taught that we should all 'cultivate our own gardens'. He wrote his most famous work, *Candide*, in three days while still reeling from the mental shock of the news from Lisbon. In this book, he has his satirical hero, Candide, involved in the Lisbon earthquake early in his adventures. Candide, the philosopher Pangloss and a brutal sailor are stranded on the shore near Lisbon as the only survivors of a shipwreck. As they approach the outskirts of the city, the earth shakes and the sea rises foaming in the harbour to dash to pieces the ships lying at anchor. The streets and squares are filled with whirling masses of flame and cinders as the houses collapse and the roofs crash down on the shattered foundations. Thirty thousand inhabitants perish in the ruins. The brutal sailor sees the disaster as a chance for plunder, Candide views it with horror as 'the day of judgment' but Pangloss sees it as an interesting scientific phenomenon for investigation. He consoles Candide and other survivors with the philosophical remark that 'all is for the best, since, if an earthquake occurs at Lisbon, then it could not have occurred in any other spot. It is impossible that things should be elsewhere than where they are; for everything is good.' Voltaire does not allow the Portuguese authorities and leading thinkers to accept this view of the earthquake – he portrays them as deciding that the best way of avoiding further destruction of the city would be to give the people an *auto da fé*, with some chosen individuals being burnt over a slow fire. Candide and Pangloss are among a group arrested for this purpose. Candide is rhythmically flogged, in time to singing, and Pangloss is hanged. Others are burnt at the stake. The next day there is another terrible earthquake.

Voltaire was not the only European thinker to be profoundly affected by the Lisbon earthquake. Goethe, who was six years old at the time, remembered later how the 'demon of fright' spread across the world. Everyone was shocked, not only by the scale of the catastrophe, but also by the apparently mindless and random nature of this so-called Act of God; with sinners and devout, young and old, rich and poor all perishing together in one awful holocaust. Nevertheless, the leaders of European thought each sought to explain the event in terms of their own prejudices. Rousseau saw it as a warning to mankind against the life lived in crowded cities, and many ordinary people, living far from Lisbon, saw it in Old Testament terms as a warning from a benevolent God regarding the growing wickedness of the world. There were even those who, blinding themselves to the facts in order to further the spread of their own dogma, could assert, like Wesley, that this warning from God had been directed 'not to the small vulgar, but to the great and learned, the rich, and honourable heathens commonly called Christians'. It was the 'small vulgar' of the Lisbon waterfront, however, who really suffered most in the catastrophe. For a short time, a serious attempt to reform society was made in England; certainly it is a fact that masquerades, which until then had been the rage of London, were abandoned for ever. It would appear that the disaster of Lisbon had as profound an effect on the minds and morals of men in the mid and late 1700s as the atomic bombs that were dropped on Japan in 1945 have had in the twentieth century. The first reports to reach Voltaire were that, within the space of 6 minutes, 15,000 men, women and children had been crushed beneath the bells and towers of Lisbon's churches, which were packed for divine service on All Saints' Day. How, some men reasoned, could one maintain belief in an all-powerful and benevolent deity who could murder thousands of innocent humans at the very moment when they are gathered to do him worship? We know now that out of a population of 275,000 living in Lisbon on 1 June 1755, between 10,000 and 15,000 lost their lives on that day. Thirty churches and most of the smaller buildings in the city were destroyed. Those which escaped total destruction by the earthquake or by its associated tsunami faced further damage from numerous fires which broke out all over the city and burned unchecked for several days. Valuable libraries, museums and monuments were destroyed, including pictures by Titian, Correggio and Rubens. Many more people died subsequently – perhaps another 15 to 30,000 – as a result of disease, poverty and other hardships indirectly caused by the earthquake, but there is no firm estimate for the total loss of life that can be attributed to the 1755 earthquake.

The world famous French volcanologist Haroun Tazieff visited southern Chile immediately after the devastating series of earthquakes that struck this remote area in May 1960 to study the overall effect of the earthquakes, collect data for the drawing of isoseismals (lines joining places where each earthquake was of equal intensity) and to observe a volcanic eruption that had broken out two days after the main shock. His account is one of the most readable and thought-provoking descriptions ever written for the general reader by a trained scientific observer: Tazieff and three companions arrived in the badly damaged town of Puerto Montt just in time to experience personally one of the numerous aftershocks of what has come to be known as 'the great Chilean earthquake of 1960'. At Puerto Montt and Valdivia the intensity of the chief shock reached XI on the Mercalli Scale and the damage to buildings, roads, railways and harbour installations was very extensive. Concrete quays were tilted, loaded trucks and warehouses thrown into the sea, ships stranded and the pier at Puerto Montt shattered into separated zigzag sections; in large areas of both towns all of the houses were destroyed. In describing their impressions of the chief earthquake responsible local people told Tazieff that they saw thick wooden pillars rippling like hanging-ropes, undulations running down pavemented roads like waves on the sea-shore and large ripples running up factory chimneys. These quite truthful impressions are nevertheless no more than illusions caused by the physiological and psychological effect of violent earth vibrations on the human sense organs and brain. Tazieff closely examined the wooden pillars, the paving stones of the road surface and the brickwork of the chimneys concerned and convinced himself that they showed none of the surface fracturing that motions of the magnitude described by the witnesses would have caused. It is clear that many of the most dramatic eyewitness accounts of ancient earth-quakes (such, for example, as those of the 1812 earthquake at Caracas, Venezuela) must include similarly truthful but equally illusionary descriptions of very large movements of the ground. Vibration and ground movement certainly do occur during an earth-quake, but not on the scale sometimes suggested by our disoriented senses.

The 1960 earthquake crisis in southern Chile began before dawn on Saturday 21 May. Just after 6 a.m. the towns of Concepcion and Chillan and the whole of the Arauco peninsula to the south were shaken by an earthquake of Mercalli intensity VIII (7.75 on the Richter Scale). In Concepcion, many buildings, church towers and old brick and stone houses collapsed and hundreds of people were crushed in their beds. Half an hour later a second shock followed as violent as the first, but as most of the rudely awakened population had by this time gathered out of doors well away from buildings the additional casualties were few. The largest of this extraordinary series of great earthquakes occurred at 3 p.m. on Sunday afternoon. If the people of the region had been in bed or indoors at the time, the casualty lists would have been enormous, for over 16 per cent of the dwellings in the region were utterly destroyed and most others severely damaged. Factories, workshops, offices, shops and schools collapsed into heaps of rubble in many places – as if the walls had suddenly ceased to exist. Fortunately, it was a sunny afternoon and numerous families were out together enjoying the unusually fine weather. In addition, two great foreshocks gave the alarm: the first of them, 15 minutes before the main 'quake, was as strong as the two shocks on the previous day; the second, less than a minute before the main 'quake, was even more violent, with a Richter magnitude of 7.8. The paroxysmal shock which followed had a Richter magnitude of 8.7 and was undoubtedly one of the greatest earthquakes ever recorded. The towns which suffered most were Valdivia, Puerto Montt and Castro (on the large island of Chiloe), in each of which the Mercalli intensity reached XI. Intensity VIII was exceeded in most of the other towns in a coastal region of Chile 375 miles long by 125 miles wide. During the eight days which followed there were three more very powerful earthquakes at different places in this region, followed by a long sequence of lesser earthshocks, spread over some weeks; one of these reached a Richter intensity of 7.0.

Soon after the greatest shock on the Sunday afternoon, people along the coast saw the ocean rise above the level of the highest tides, then suddenly retreat very rapidly far beyond the lowest low-water tide mark. The population of coastal Chile are only too familiar with the phenomenon of seismic sea waves and most of them fled to the hills immediately they saw this happen, otherwise the death toll from the ensuing tsunami would have been very much greater. At Corral, a little seaport 10 miles from Valdivia, the waters of the estuary became violently choppy during the earthquake and sand was thrown up rhythmically from the shoals, appearing like the backs of surfacing whales. Following the cessation of the vibrations, the sea began to rise smoothly but with increasing rapidity, until it had flooded through the streets of the town, reaching perhaps 12–15 feet above its normal level. It maintained the high level which it reached by four o'clock for five minutes. Three ships at anchor broke free from their moorings and were later swept away, for at 4.10 p.m. the sea began to withdraw at great speed accompanied by a horrible sucking noise, carrying one ship, the 3000-ton *Santiago*, right over the concrete mole. Foolishly, many of the men of Corral rushed back down the nearby hillside into the town to try to salvage some of their belongings as the waters retreated and were thus caught when the second wave, a 26-foot high tsunami travelling at 125 m.p.h. struck the little town at 4.20 p.m., completely destroying all the buildings. At the foot of the hill on which the women and children were huddled the remains of 800 houses were heaped up like so much matchwood, together with the bodies of those men who had returned to the town. Fishing boats at sea were swallowed without a ripple by the wall of ocean water as it rushed towards the shore. The sea remained high for 10–15 minutes, then retreated once more with the same dreadful sucking noise. An hour later a third, even higher, wave approached the coast: 30 or 35 feet high, it was seen from its first appearance far out in the ocean sweeping in remorselessly at close to 60 m.p.h. Smaller waves followed but this third wave was the last of the destructive tsunamis. In Corral, there was nothing more it could destroy.

Inland, many valleys were blocked by landslides and the danger of disastrous flooding was present for several weeks. The collapse of these temporary natural dams before the weight of water trapped behind them was awaited with some anxiety. For two months frantic efforts were made to drain away the enormous volume of excess water that was accumulating in Lake Rinihue behind a great landslip in the river valley just below the lake outlet. Valdivia had been destroyed once before, in 1575, by a similar event, and the authorities were naturally apprehensive. A great canal was cut through the landslip and the dangerous waters allowed to escape under human control. Further inland, in the cordillera and foothills of the Andes, thousands of landslides and avalanches were triggered by the great earthquake. Studying some of the numerous landslides that had devastated inhabited valleys in the foothills, burying hundreds of innocent farming families beneath their own cultivated top soil, Tazieff and his companions were struck by the unexpectedly great horizontal distances covered by these landslides. It became clear that under the influence of intense ground vibration, the movement of large masses of soil and debris dislodged from the sides of quite gentle hills can be maintained (i.e. they remain fluidized) for a much longer time than under normal static conditions and, as a result, the landslides had travelled out across the flat floors of valleys for distances up to five or six times the height of the hillside scar from which they were derived. We now know that seismic landslides and avalanches can move faster and for far greater distances than those which occur under normal conditions in mountainous regions and thus can be considerably more destructive and are even more terrifying than those relentless mountain killers with which most people are already familiar. In an attempt to reach the eruptive fissure on the north-west flank of the volcano Puyehue that had spewed out andesite lavas two days after the earthquake, Tazieff and his party travelled up the valley of the Rio Gogol. Here, in the high cordillera, they saw monstrous landslides that had either removed entire mountainsides or, beginning as smaller slips on the higher slopes, had torn on down through the forested lower slopes as terrible rock avalanches. The inhabitants talked of the 'running hills' that had buried a hundred roadmakers in their camp east of Lake Rupanco, hurled the hotel and staff at Rupanco Spa into the lake and buried numerous farmers instantly under thousands of tons of cascading earth and stones.

At many places in the region devastated by the 1960 earthquakes the ground was fissured and a variety of ground subsidence or uplift had occurred. Permanent flooding by either river or sea water was observed to be extensive for many miles of the coastline north and south of Valdivia, around Maullin and along the east of the island of Chilloe. Even after careful examination of the evidence Tazieff found it hard to appreciate the fact that an immense stretch of land 18 miles wide by 300 miles long had sunk between 6 and 10 feet in 10 seconds or so. In general, buildings situated on firm hard-rock foundations survived better than those on alluvium, soft valley fills, coastal plains or made ground. Buildings erected since 1939 that complied to special design regulations intended to help them withstand the effects of earthquake did take the strain well. Old, badly or shoddily constructed buildings failed everywhere. Much damage was caused in well-designed and constructed buildings, however, due to the fall of heavy chandeliers and the violent crashing about of loose items of furniture.

The true scale of this immense event can only be seen if it is imagined to have taken place nearer home. The sunken strip of land would stretch from Canberra to Melbourne if it were in Australia, from Paris to Bordeaux in France, from Glasgow to Oxford in Britain or from Memphis to New Orleans in America.

It is clear that human experience of catastrophic earthquakes is very dependent on the environment in which they occur. Away from big cities and other populated areas the effects and dangers can be very different from those experienced by the population of Lisbon in 1755, San Francisco in 1906, southern Chile in 1960 or of Tientsin in 1976. On the afternoon of 15 August 1950, for example, seismographs around the world registered an exceptionally severe tectonic earthquake in the Himalayan mountain chain just inside Assam, near the borders of India and Tibet. (The magnitude recorded was, in fact, the largest since instrumental observations began.) It is likely that this great Assam earthquake was related to massive rock rupture under stress at a depth of about eight miles, and subsequent earth movements along major faults in an area where there is a sharp right angle bend in the mountain chain. Aftershocks, some of considerable magnitude, continued for over two months as the mountains settled and readjusted themselves after their first violent convulsions. The intense earth shaking and the subsequent floods affected an area of about 70,000 square miles along the Brahmaputra River from India to north-eastern Tibet, leaving very few buildings undamaged. Ernest Tillotson reported that most of the countryside on each bank of the river sank, in some places between five and six feet. Roads were twisted, cut up, breached and destroyed. Agricultural soil was upturned and fields flooded, causing havoc in the rice fields. It was too late to plant again that season. There were multitudes of dead fish. The dam in the main Brahmaputra River north of Pasighat broke, causing serious flooding. The Subabsiri River was dammed by fallen rock where the river comes through a gorge with cliffs 1000 feet high; when this natural dam broke, the sight must have been awe-inspiring. Damage to property was extensive but the death toll was low – between 1500–2000 persons – because there were no major cities in the worst affected area. The British naturalist, F. Kingdon-Ward, was camping in northern Assam at the time of the earthquake and his eye-witness description is one of the most dramatic accounts ever written of what it is like to be at the centre of a severe earthquake in the mountains.

On 15 August 1950, Kingdon-Ward and his wife were camped in the village of Rima, on the left bank of the Lohit River, at an altitude of 5000 feet. (Rima is not really the name of a village, but rather of a small district containing five or six villages of rude timber houses and granaries, with a total population of a few hundred people.) All around, the mountains rise very steeply from the flat terraces, a cluster of snow peaks on the Burma frontier, more than 19,000 feet high lying within the 25 mile radius; but only 4 miles from the river they are already 15,000 feet high. About 8 p.m. local time, roughly three-quarters of an hour after dark on a hot, close night with the stars shining brightly upon the arid gorge, Kingdon-Ward was seated near the entrance to his tent, writing his diary. His wife was in bed, the two servants in their own tent. Suddenly, after the faintest tremor, there came an appalling noise, and the earth began to shudder violently; Kingdon-Ward jumped up and looked out of the tent; the outlines of the landscape, visible against the starry sky, were blurred – every ridge and tree fuzzy as though it were rapidly moving up and down; 15 or 20 seconds passed before he realized that an earthquake had started. His wife shouted 'Earthquake' before he did, and leapt out of bed. Together they rushed outside, seizing the oil lantern for fear the tent would catch fire, and placing it on the ground. Immediately they were thrown to the ground; the lantern, too was knocked over, and went out instantly.

Afterwards they found it difficult to recollect their emotions during the four or five minutes the shock lasted; but the first feeling of bewilderment – an incredulous astonishment that these solid-looking hills were in the grip of a force which shook them as a terrier shakes a rat – soon gave way to stark terror! Yet, lying side by side on the sandbank, they spoke quite calmly together and to the two Sherpa boys who, having already been thrown down twice, were lying close to them.

The earthquake was now well under way and to the four travellers lying together it felt as though a powerful ram were hitting against the earth beneath them with the persistence of a man beating a drum. Kingdon-Ward had exactly the sensation that a thin crust at the bottom of the river basin, on which they lay, was breaking up like an ice floe and that they were all going down together through an immense hole into the interior of the Earth. The din was terrible; but it was difficult to separate the noise made by the earthquake itself from the roar of the avalanches pouring down on all sides into the basin.

Gradually the crash of falling rocks became more distinct, the frightful hammer blows weakened, the vibration grew less, and presently they knew that the main shock was over. The end of the earthquake was, however, very clearly marked by a noise, or series of noises, which had nothing to do with falling rocks. From high in the sky to the north-west (as it seemed) came a quick succession of short, sharp explosions, like 'ack-ack' shells. These noises were 'rock-bursts', the explosive failure of large rock bodies under the stresses imposed on them by the earth movements. It was the 'cease-fire'.

After about half an hour Kingdon-Ward and his wife returned to their tent. He noticed that the travelling clock on the table and his watch were both going; the altimeter still registered 5000 feet exactly and the thermometer outside under the fly showed 73°F. Nothing inside the tent was disturbed, except a glass of water which had been upset. The tents were pitched on a small sandbank about 600 yards from the high bank of the La Ti torrent, rather more from the Lohit River. The nearest mountainside was 300–400 yards to the east and, as it happened, that particular face did not slip badly; at any rate, no boulders reached the tents or even the nearby village. They could not have selected a safer site in an earthquake! Within two hours the air was so thick with dust that every star was hidden; they breathed dust, it gritted their teeth, filled their eyes. Within a week every leaf of every tree was caked with dust. Violent tremors continued all night and twice the small party rushed outside – though the tents were perfectly safe.

As soon as it was light, they dressed and went out to see what had happened. In the village every house had lost its roof of wooden slates and every lean-to had been thrown down. The main irrigation channel was blocked and the paddy fields with their standing rice would soon be dry. Long fissures cut across the stony fields, running for the most part parallel with the river banks, past or present. In some places numerous fissures lay close together, elsewhere they were far apart – it appeared to depend on the nature of the soil, whether sandy or clayey. These cracks were rarely more than a few inches wide and two to four feet deep. Every well-trodden path was cracked throughout its length. Here and there a small block of land had sunk bodily. The La Ti torrent was an extraordinary sight. The previous day the sparkling water, of a beautiful blue-green tint, revealed every boulder as it flowed swiftly down its rocky bed. After the earthquake, it was dark coffee-coloured, but so thick with foam that it looked like café au lait. It had risen nearly four feet, and the muffled grinding of submerged boulders sounded ominous. The water smelt unpleasantly of mud, but Kingdon-Ward could detect nothing sulphurous about it.

Day after day, night after night, tremors followed each other. Many of them were severe and were always preceded by a short roar, like a distant thunder clap. The weather was sultry, the maximum shade temperature often more than 95°F, until a wet spell between 23 and 27 August brought it down; the minimum was usually about 65°F. On fine days a breeze blew from the south, but in the Rima basin it was so charged with fine dust and so hot that it brought no relief. Every day by midday, as the rock avalanches began to fall, a vast dust cloud hid the sun or dimmed it until it looked like a copper plate. Dust mingled with the gathering thunderclouds and visibility was reduced to a mile or two. No one was killed in Rima, but cattle and pigs were injured or killed by falling timber; in the river, great numbers of fishes were killed, although few were taken from those raging waters. The indifference, or fatalism, of the local population (nominally Buddhist) was superb. The day after the earthquake they went about their work in the fields as though nothing had happened, yet many of them seemed to think that there would be a repetition of the major shock and that next time they might all be killed. Kingdon-Ward and his party could find no one who recollected a previous earthquake. With their houses exposed to the weather, they slept in temporary shelters made out of boards, mending their roofs in the evenings: the village at least looked normal again after a fortnight.

As for the mountains which enclosed the basin, they had everywhere been badly mauled. Wide belts had been ripped off, carrying trees and rocks; whole cliffs had crashed down, deep wounds scored; and everywhere rocks continued to cascade down hundreds of gullies. The damage done in the main Lohit valley was bad enough; but that done in the tributary valleys, where every stream had to break through a narrow gorge thousands of feet deep, was infinitely worse. The destruction extended to the tops of the main ranges – 15–16,000 feet above sea level. No wonder the mountain torrents began to flow intermittently as the gorges became blocked, followed later by the breaking of a dam; whereupon a wall of water 20 feet high would roar down the gully, carrying everything before it and leaving a trail of evil-smelling mud. The Kingdon-Wards, their servants and the villagers had survived one of the greatest earthquakes in history!

Earthshock

ake shook the area.
lative elections was
second ballot was
preoccupations
on to discour-
as to suggest
le leaving:
possible.
of the
the
ck

of fire.
he sun.
s are sinking.
heavens are falling.
ne hole of the pit.
, and the wolf range free.

Icelandic Sagas

In 1658, the French and the Caribs were fighting fiercely in the north of the Caribbean island of Martinique. Eventually, after a bitter struggle during which they suffered ruthless decimation by their enemies, the proud Caribs first blinded themselves, then threw themselves from the top of a high cliff in order to avoid the disgrace of capture. This cliff is part of a hill known as the 'Caribs Tomb' and lies a few miles along the coast from the town of St Pierre. Before the Caribs made this sacrifice they invoked their gods and declared: 'The Fire Mountain will avenge us.'

On 8 May 1902 Mont Pelée, the Fire Mountain, did indeed avenge them. In February 1902 the inhabitants of Le Prêcheur, a small fishing village 5 miles north-west of St Pierre, had noticed sulphurous smells which gradually became stronger. Later in the month, birds were choked to death by the fumes and their corpses were found scattered about the hillside. Even horses became restless when led into certain areas, where fumes were seeping to the surface. These manifestations continued and increased during April until the 23rd when, at eight in the morning, a strong earthquake was felt in St Pierre and neighbouring villages. On the following day an enormous 'bump' was heard, followed by muffled noises and rumblings which appeared to emanate from the bowels of the Earth. On the morning of the 25th it was unusually cloudy and the sky became dark, as though there was an eclipse of the Sun; then a sudden report rang out, like a cannon shot, and the sky was ablaze. For several hours incandescent fountains sent showers of fine ash cascading down over the villages and towns on the lower slopes of Mont Pelée, causing a great deal of fright but little damage. At ten in the evening another earthqu

On Sunday 27 April, the first ballot for the legis
held without an absolute majority being gained; a
fixed for 11 May. It was later said that these politica
were the main reason for the Town Authorities' decis
age evacuation. Indeed, some reports have gone so far
that road blocks were set up in St Pierre to prevent peop
the politicians wanted, quite naturally, the largest majority
Even the sudden appearance of swollen rivers on the side
mountain was disregarded.

May began with further earthquakes and increased ash falls
Le Prêcheur area was covered with a layer of ash several inches th
– and people from the outlying villages began crowding into St Pierr
On 2 May the local newspaper, *The Colonies*, actually advertised an
excursion to view the eruption:

> The great excursion to Mont Pelée, organized by the Gymnastic and
> Shooting Society, is to take place next Sunday, 4 May. Those who have
> never been up there to enjoy the panorama that greets the astonished
> visitors' eyes at an altitude of 4590 feet; and those who wish to see the
> still gaping hole through which, these last days, broke forth the thick
> clouds of smoke that filled the inhabitants of the upper areas of Le
> Prêcheur and Saint-Philomene with fright, will have to take advantage
> of this nice opportunity. Weather permitting, the excursionists will
> spend a day which they will long remember with pleasure.

At 11.30 p.m. on 2 May the sleeping town was roused by a series of
loud detonations, while an enormous column of black ash, streaked
with incandescent material, burst out and showered the countryside
with pumice and ashes that the wind blew as far as Fort de France, 20
miles away. That night, Holy Communion was received by a panic-
stricken crowd. The volcano was by now erupting vigorously and
hurling large blocks of rock over 1½ miles from the vent. Everybody
in St Pierre was coughing and suffocating in a thick murky fog which
prevented the tramway from running; the town was soon covered
with fine ash, and on 3 May people crowded into the Cathedral and
asked for absolution. The great excursion that had been arranged for
the following day was cancelled.

On Sunday the 4th, a new crater burst out on the side of the volcano
and the Rivière Blanche, which had been in flood previously, now ran
dry. Then one of the craters spat out a terrible gush of mud, and the
Rivière Blanche became a furious torrent of black sludge, carrying
enormous boulders along with a deafening roar; it had to pass the
Guerin distillery, near the mouth of the river, to reach the sea, and it

simply bulldozed the distillery buildings into the sea, killing 23 people and creating huge waves which broke upon the beach at St Pierre, destroying boats and flooding the Place Bertin. The volcanic violence continued unabated throughout the night and the following day, with torrential mudflows crashing down river valleys and ash falls covering the countryside.

On Wednesday 7th, in what should have been a salutary lesson to the inhabitants of St Pierre, an eruption of Soufrière volcano, on the nearby island of St Vincent, sent an incandescent cloud of gas and pulverized rock searing through the forest, killing 1565 people in the Georgetown area. Yet, on the same day, during a temporary lull in the eruption of Mont Pelée, this communiqué was issued in Martinique: 'According to exterior signs, the intensity of the eruption is decidedly declining. The height of the column of ashes, which, last Sunday night, reached about 5700 yards, only reached 2600 yards this morning. The outflow of steaming mud in the valleys of the Rivière Blanche no longer goes as far as the sea. Many tourists have made for the crater.' A scientific mission was entrusted by the government with the study of the characteristics of the Mont Pelée eruption. Led by Lieutenant-Colonel Gerbault, a technician and artillery director, the mission consisted of a pharmaceutist-officer in the colonial army, a government civil engineer and two natural science teachers at the lycée of St Pierre. Later, one of the teachers issued another communiqué: 'In my opinion, our Mont Pelée does not endanger the town of St Pierre more than Vesuvius endangers Naples.' Naples is 6 miles from Vesuvius, however, and St Pierre only 4½ miles from Pelée and, understandably, in spite of the official advice they were receiving, panic broke out among the inhabitants in the afternoon and barricades are reported to have been erected.

At this time the Governor decided to try to calm the populace by visiting St Pierre; after all, the elections were drawing close. He never left. That night, the detonations continued, and much ash was mixed with the torrential rain that fell continuously upon the town. About 4.00 a.m. on 8 May there was another brief lull in activity, but the panic-stricken people rushed about gathering up their possessions ready to leave at daybreak; others stormed the harbour trying to get to the boats. The situation was now uncontrollable.

At dawn the sky was clear over a town washed clean by torrential rains. It was Ascension Day, and the Angelus bell was tolling. People crowded into church. Time, however, was slipping away. At 7.50 a.m. there was a loud explosion high on the west flank of the volcano.

An eyewitness out at sea reported that, as the explosion sounded, the mountain appeared to split open from top to bottom. While one black cloud gathered above the peak, another projected itself at hurricane speed through a notch in the side of the volcano and hurled itself upon St Pierre (Plate 13). By 7.53 a.m. it was all over. Nothing survived. The fiery blast, like that of an atomic bomb, had flattened the town, and everything in the path of the terrible *nuée ardente* (burned cloud) was incinerated. The toll was horrific – 28,000 dead –

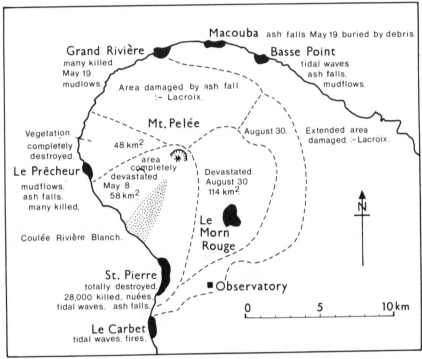

Fig. 20   Volcanic hazard: Mont Pelée. Hazard map for the northern part of Martinique, Lesser Antilles, illustrating the extent and magnitude of damage caused by the catastrophic eruption of 1902. A similar eruption today could cause as much death and destruction as shown above if the population is not evacuated before the eruption climaxes.

and the 'Little Paris of the Caribbean' lay in ruins (Plates 14 and 15). Enormous pieces of the Cathedral had been hurled many yards and one of its towers appeared to have been ripped apart in the middle. Offshore, the hulks of boats were burning, slowly rocking on the almost calm sea. On the shore lay the debris of cargo: bales, barrels, beams and shipframes.

When the first would-be rescuers reached the town, the air reeked with the stench of putrid flesh mixed with the acrid smell of burnt bodies (Plate 16). There was only one survivor from the central zone of the holocaust. Augustus Cyparis, a native of Le Prêcheur, had been sentenced to prison for assault and battery in a street brawl. He had almost served his time, when, learning of a local fête, he managed to escape from custody while being transported to work in the town. At the end of the festivities Cyparis gave himself up and was promptly sentenced to eight days in the dungeon. The punishment saved Cyparis from certain death. Three days after the tragedy, when he thought that the end of the world had come, he was found in the dungeon, his bloated body covered with numerous burns. (Later he was able to put his scars to good use by exhibiting them in a travelling circus.) His dungeon can still be seen behind the ruins of the theatre in St Pierre: it is built of large stone blocks, rather like a small bomb shelter, with only two small windows.

In the St Pierre tragedy, the eruption itself was directly responsible for the high death toll. Nevertheless, secondary phenomena induced by primary volcanic action can have equally far-reaching effects, as in the case of the mighty eruption of Krakatoa volcano in 1883, when giant tsunamis destroyed shore-line villages in Indonesia causing tremendous loss of life, and the widespread epidemic and famine which followed added a further dimension to the catastrophe.

The tremendous eruption of Krakatoa is one of the most informative examples of a volcanic eruption in historic times; during its violently explosive outburst the island of that name all but disappeared, many thousands of people died, and the world suddenly became aware of the possible extent of volcanic violence. Krakatoa lies in the centre of the Sunda Straits, at the western extremity of Java, approximately half way between the Java and Sumatra coasts at a distance of about 20 miles from the surrounding shores. These narrow straits were one of the most important Pacific shipping lanes during the nineteenth century, being part of the route from Batavia (now Djakarta) to Europe via the Indian Ocean. Indeed, it was because large numbers of ships used this popular route that we know so much about the fateful events of 1883.

The details of the shorelines were well known to the naval hydrographers of that time, since both the British and the Dutch were engaged in active trade with that part of Indonesia and beyond. The name Krakatoa is believed to be a corruption of Rakata, a 2667-foot high volcano which, in 1883, stood at the southern end of the main island in the group. Just north of Rakata lay Danan, another volcano, 1476 feet high and to the north-west of Danan the lower Perboewetan volcano, height 394 feet. North-east of the main island was Lang Island and north-west lay Verlaten Island. Half a mile west of Lang Island was another small island with the curious name of Polish Hat.

Of the three obvious volcanoes, Rakata appeared to be the youngest, with fresh-looking lavas on its flanks. Eruptions of Krakatoa had been reported between 1680 and 1681 and it is possible that a large part of Rakata cone was built at that time. The other cones were associated with more ancient lavas and nothing is known of their past history. All the islands were uninhabited but were visited periodically by people from the neighbouring mainland in search of the timber which grew particularly well in the steamy tropical forests that cloaked the islands. Krakatoa was, in fact, an exceptionally beautiful tropical island – a Garden of Eden – much admired by the passengers of passing ships. It very soon turned out to be a Pit of Hell.

During the 1870s the area around the Sunda Straits was shaken by several earthquakes, but no one living there took any notice. After all, earth tremors were not uncommon in this region. (It is interesting that Pompeii was also shaken by a sequence of earthquakes, beginning 16 years before its total destruction in AD 79, and that there too the warning signs were ignored.) As is common with volcanic eruptions, the frequency and violence of the 'quakes in the Sunda Straits increased until, on 1 September 1880, a particularly large tremor caused structural damage to a lighthouse on the Java coast. Even after this clearly premonitory shock, people went about their business quite normally. The seismic activity continued sporadically until 1883, but was regarded as no more than a nuisance. On 20 May 1883, however, Krakatoa suddenly burst into activity. Loud explosions were distinctly heard for several hours as far as 100 miles away, where doors, windows and other loose objects rattled in the air pressure and ground vibration waves caused by the blasts. Compass needles were agitated on ships. The next day a huge column of steam six miles high was observed issuing over Krakatoa island, while the wind carried fine ash up to 300 miles away from the island. On the 22nd a dome-shaped mass of vapour appeared as great quantities of pumice fragments were hurled out of the active vent. The eruption cloud continued to soar skywards and at night was rent by fantastic lightning displays. On the 23rd much pumice was seen floating in the sea but at this time there was no hazard to shipping. Both the crews and passengers of the ships that used the Sunda Straits that week no doubt viewed this display of the forces of nature with unconcerned curiosity. Up to this point the activity was very similar, it appears, to that which had occurred in 1680 and 1681, with the eruption column rising to 36,000 feet and volcanic dust being carried downwind for at least 300 miles.

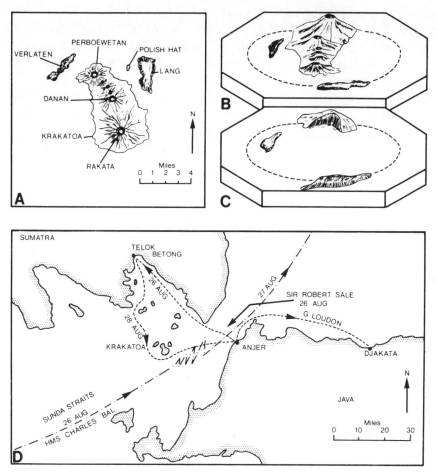

Fig. 21   During the cataclysmic eruption of the volcano Krakatoa in the summer of 1883, much of the former volcanic island vanished in a great caldera subsidence. Map (A) shows the island group as it was before 1883 and the two block diagrams (B) and (C) illustrate how the volcano would have appeared from high to the north immediately before and after the great eruption. Map (D) indicates the movements of the three ships, HMS *Charles Bal*, the *Gouvenour-Generaal Loudon* and the *Sir Robert Sale* in the Sunda Straits between 26 and 28 August 1883. The greatest loss of human life was not from the direct effects of the eruption, but from the associated tsunamis that inundated surrounding coasts.

On 27 May there was a lull in activity. This prompted a group of sightseers from Batavia to charter a boat and set sail for the island to view the volcano at close quarters. Their trip, however foolhardy, was of considerable scientific value for it gave us the last detailed description of the island before it vanished in the cataclysm that was to come. They found that strong explosions were coming from Perboewetan cone, which was in eruption, while the rest of Krakatoa and the neighbouring islands were covered with a thick layer of fine white dust.

On Verlaten Island and in northern Krakatoa, trees had been stripped of their leaves and branches by falling pumice and their withered stumps stood like so many lonely sentinels amidst the desolation. Going ashore, the visitors stepped into ankle-deep ash and mud and, although irregular explosions from Perboewetan were hurling blocks of pumice 650 feet into the sky every five to eight minutes, they continued their observations and were able to see the bright glow of the lava in the crater. The main crater of Perboewetan was then roughly 3300 feet across and 160 feet deep. At the bottom was an open conduit: a vast column of steam was rising from this and ascending to over 9800 feet with a great roar.

Activity on Krakatoa declined further after this visit until, in the middle of June, it appeared that the eruption might be nearly over. That this was not the case was revealed on the 19th, when a great plume of ash and steam shot up, this time not only from Perboewetan, but also from Danan volcano in the centre of the island. The strong explosions which accompanied this recrudescence of violent volcanism, were, no doubt, responsible for the disappearance of parts of Perboewetan.

The complacency of the populace on Java and Sumatra during July was truly remarkable. Whole areas were being shaken by explosions and rocked by earthquakes, yet the authorities and the people apparently did nothing. By early August another major vent had opened on the island of Krakatoa. On the 11th the island was examined from the sea by the Dutch Government Surveyor, Captain Ferzenaar, and was seen to be now completely devoid of vegetation. One of the active vents was still on the site of the original Perboewetan crater, the second was at the foot of Danan volcano in the centre of the island, and the third was the main cone of Rakata. Steam and ash were also issuing from a dozen or so minor openings all over the island and this pall of fume and ash made any closer observations impossible. Krakatoa had now been in eruption for 84 days but had still not entered the most violent, or *plinian*, phase of its eruption.

Practically all the details of subsequent events in the history of this eruption were painstakingly compiled after the disaster from observations made by those who chose to respond to the call for information. Particularly useful were the log books kept by captains of ships which had sailed through the Straits of Sunda during the last week in August 1883, for here details of what happened were recorded with the accuracy for which ships' logs are renowned.

From 23 to 26 August there was a marked, though gradual, increase in the violence of the eruption, with the volcano hurling out a black mass of ash and pumice which rose to a height of over 17 miles – 90,000 feet – while every few minutes the shock of explosions and earthquakes caused buildings to tremble. On the 26th Krakatoa entered the plinian phase of its 1883 eruption. From 2.00 p.m. sharp, detonations were occurring every 10 minutes which could be heard 100 miles away. Strong air vibrations were experienced by 3.00 p.m. and heard 150 miles away, while all the time the dark plume of ash and gas continued to rise to 17 miles. Still the violence increased, and by 5.00 p.m. the explosions were being heard all over Java, a distance of up to 700 miles. In Batavia, 100 miles away, the sound was deafening; pictures, windows, doors and chandeliers rattled. Large pumice fragments fell 10 miles from the volcano and incandescent ash and lava could be seen at times coming from the vents. Violent electrical storms lit up the sky by night while the eruption continued unabated.

On the same day – 26 August – a British naval vessel, HMS *Charles Bal* was sailing through the Straits of Sunda en route for Hong Kong. In the early afternoon the ship came within 10 miles of Krakatoa. Captain Watson recorded in his log that deafening explosions, sounding like heavy artillery and accompanied by loud crackling sounds, repeatedly rent the atmosphere – the crackling was probably large volcanic bombs of hot lava, highly charged with gas, exploding in the sky. By 5.00 p.m. conditions were so severe that Captain Watson, fearing the worst, gave orders to shorten sail on his ship. During the next hour a hail of large pumice fragments, some as big as pumpkins and still hot, fell on the decks of the *Charles Bal*. With sailing conditions so bad it became necessary for the vessel to tack to and fro south-east of Krakatoa, now nearly 13 miles distant; visibility was almost non-existent at times, but Captain Watson found it possible to use the glow of the volcano as a lighthouse and set his new course by it. There was a great deal of static electricity in the atmosphere, generated by the movement of tiny particles of rock and droplets of water from the steam, which caused spectacular brush discharges to take place from the masts and rigging of the ship – a phenomenon known as St Elmo's Fire. The crew were greatly alarmed by this and feared that the fire would fall onto the deck and burn it.

A local Indonesian vessel – the *Gouvenor-Generaal Loudon* – was also in the Straits and experienced conditions similar to those encountered by the *Charles Bal*. It spent the night of 26 August anchored at Telok Betong in southern Sumatra. Her captain attempted to sail for Anjer on the morning of the 27th but was prevented from doing so by the increasing violence of the eruption. A third ship, the British *Sir*

Robert Sale, entered the narrow eastern end of the Sunda Straits on the 26th, intending to sail past Krakatoa on the 27th, but was quite unable to proceed.

On 27 August, after a night of terror everywhere close to the Straits, with explosions taking place every second from midnight until 4.00 a.m. in an almost continuous roar, the eruption reached its climax. At 5.30 a.m., possibly in response to sea water gaining access to the vent, there occurred an enormous explosion (its time was recorded accurately by the barograph at the Batavia gasworks). This, however, was only the beginning, for at 6.44 a.m. there was another great explosion and at 10.02 a.m. yet another, the largest of them all. The ash plume was now rising to heights of 50 miles and ash was raining down over an area of 300,000 square miles. The third explosion caused an extraordinary increase in atmospheric pressure, which was again measured by the Batavia gasworks barograph, while in the town windows were broken and walls were cracked. (The gigantic eruption of the volcano Bezymianni in Kamchatka, in 1956, resulted in similar atmospheric pressure waves which caused pain in the ears at a distance of 31 miles.) Further cataclysmic explosions at Krakatoa took place at 10.52 a.m. and 4.34 p.m., and it was during these culminating explosions that the island of Krakatoa disappeared. The detonations which accompanied the five cataclysmic explosions were heard over a large area of the Earth: in Singapore and Australia the noise was loud enough to sound like nearby gunfire. The furthest place to which the sound carried was Rodriguez Island, 2983 miles away, on the other side of the Indian Ocean!

The gigantic eruption of Krakatoa caused little direct damage to property or loss of human life, because the island was uninhabited. It was the associated tsunamis, generated by the explosions and probably by the wholesale foundering of Krakatoa Island into the evacuated magma chamber below, which caused the catastrophic devastation to the shores of Java and Sumatra. The first tsunami hit the coast at approximately 5 p.m. on 26 August and subsequent waves caused further devastation throughout the night. At 7 a.m. on the following day, in Telok Betong, the gunboat *Berouw* was struck by a series of 87-foot high waves which carried her inland for over a mile. In the Sunda Straits the tsunamis were between 65 and 100 feet high and at Merak over 130 feet. A few miles south, at Pepper Bay, tsunamis swept inland for 10 miles causing terrible destruction. This was the scene all along the coastal areas of Java and Sumatra, where entire villages were swept away together with their populations. Anjer seaport was obliterated. In all, some 36,417 people died as a result of these killer waves, and thus, indirectly, as a consequence of the plinian eruption of Krakatoa volcano.

As the ash from the eruption slowly settled through the atmosphere it turned day into night in many areas. In Batavia ash began to fall at 10.39 a.m. on 27 August, 29 hours after the first gigantic explosion – an air drift of 3 m.p.h. By 12.20 p.m. the fall was so heavy that the sun was blotted out in the stygian gloom. This lasted until 1.00 p.m. when light, albeit of a sickly yellow colour, returned. By 3.00 p.m. the ash fall ceased. Batavia was lucky, for nearer to the volcano total darkness persisted for 2½ days. During this appalling fall of ash the crews of HMS *Charles Bal* and the other ships were forced to work continuously in order to clear the decks and rigging, lest the additional weight should cause the ships to capsize. At times, ash was accumulating on the deck of the *Charles Bal* as fast as 6 inches in 10 minutes. Further volcanic activity was resumed at 7.00 p.m. on 27 August, but by one hour to midnight it was obviously on the wane and at 2.30 p.m. on the 28th the eruption ended, after 121 days of activity.

The disruption to the social organization of Java and Sumatra was severe. Shipping found an additional hazard after the eruption in the form of immense rafts of floating pumice 10 feet thick and scores of miles in length. A survey by the Royal Dutch Navy later revealed the extent of the damage and showed that Polish Hat and the whole of the northern half of Krakatoa Island had vanished. All that remained of the main island was an arcuate cliff composed of part of Rakata cone. Both Danan and Perboewetan had vanished too, and in their place was an enormous undersea crater over 4 miles across and 910 feet deep, connected with a second submarine crater which was 3 miles wide and 400 feet deep.

Krakatoa Island had gone, but what had happened to it? One of the most popular hypotheses was that it had been simply blown to pieces; but careful investigation of the old records and of the composition and volume of the volcanic debris deposited on the other islands of the region reveals that this was not so. The amount of solid rock debris derived from the previously existing islands of the Krakatoa group amounts to no more than 5 per cent of the deposits. When the minimum total volume of material erupted is calculated and then converted into the equivalent volume of dense rock – a standard procedure in volcanology when trying to link pyroclastic material (dust, ash, pumice, bombs) thrown out by large volcanic eruptions with the subsidence caused by the eruption – the amount, 53,000,000,000 cubic yards, is just sufficient to fill the submarine craters caused by the eruption. This indicates that it was the zone beneath the volcanoes that was blown away during the eruption: in other words a subterranean magma chamber, filled with hot liquid lava before the eruption, had been almost completely evacuated into the air through the explosive vents. During the climax of the outburst, gas and magma were being released at a tremendous rate and this rapid evacuation of the magma chamber removed much of the support for the foundations of the island's superstructure. Gravity did the rest, the volcano simply foundering by stages into its own nearly empty magma chamber. It is highly probable that, as this took place, vast quantities of sea water gained access into hot cavities in the crust, some still containing hot magma, and that the virtually instantaneous conversion of this sea water into superheated steam caused some of the tremendous gas blast explosions that characterized the eruption. Many of the tsunamis may also be correlated directly with these events. The area continued to subside for several years after the eruption had ended, as compaction and bedding-in of the huge crater, more correctly called a *caldera*, took place.

Fine volcanic dust carried into the atmosphere during the eruption was borne away to the west by the upper atmospheric winds. Two days later the dust plume was over Africa and 15 days later it had circumnavigated the world, causing a distinct haze in the sky throughout the tropics that gradually spread north and south, reaching as far as Iceland by 30 November. This dust haze caused a variety of optical effects, including magnificent sunsets, coloration of the Sun and Moon and the appearance of coronas. Green, blue and pink Suns and Moons were seen all over the world! More significantly, when the dust pall first arrived over Europe, scientists noted that the radiant energy being received from the Sun at ground level dropped by 20 per cent. Readings about 10 per cent below normal were recorded for several months. In all, it took more than three years for the atmosphere to be cleansed of the fine dust from Krakatoa.

The 1883 eruption of Krakatoa made a great impression on the scientists of the Victorian world. Never before had so great an instantaneous release of natural volcanic energy at one locality been observed and recorded. Indeed, the man-made atomic explosions of recent years are still far smaller than the natural detonations of 1883, and the death toll from the associated tsunamis was also the highest of which we have definite knowledge.

Nevertheless, even the eruption of Krakatoa pales to insignificance when compared with some of the cataclysmic volcanic events of the past, the evidence for which is contained within the geological history

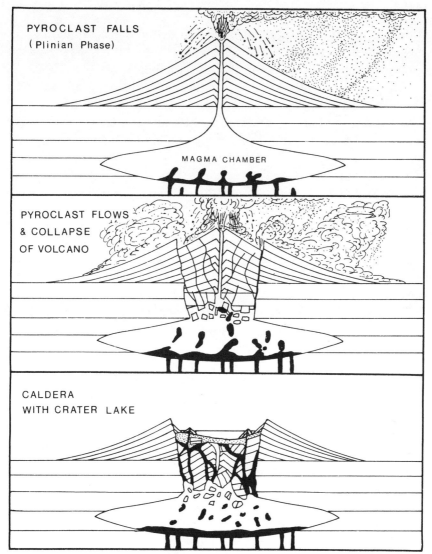

PYROCLAST FALLS
(Plinian Phase)

MAGMA CHAMBER

PYROCLAST FLOWS
& COLLAPSE
OF VOLCANO

CALDERA
WITH CRATER LAKE

Fig. 22 Caldera formation: the upper diagram illustrates the first stage in caldera formation. During the plinian phase a steadily reverberating eruption explosively discharges huge volumes of pyroclasts, which fall and mantle the surrounding countryside under a blanket of ash. Near the vent large dense blocks are unaffected by wind drift and pursue ballistic trajectories to land around the vent. The volcano is underlain by a laterally extensive magma chamber which is thick enough to prevent basaltic feeder dykes (black) reaching the surface. Instead, the basaltic feeders supply energy to the magma in the form of heat, and assist in topping up this huge reservoir after each eruption.

As the gas pressure falls during an eruptive episode the eruption column may collapse by bulk subsidence. More likely, however, the low gas pressure may produce a foam lava which will shatter into a dense cloud of red hot fluidized particles. Either process will produce a pyroclast flow. This stage is often associated with fracturing and progressive foundering of the volcano superstructure into the partly evacuated magma chamber.

In the lower diagram the collapse is complete. The volcanic heat and gas energy is spent and the magma chamber begins the slow process of refilling and absorbing the foundered cone. The circular depression at the surface is known as a caldera and may contain a lake, as well as one or more post collapse cones.

of the Earth. For example, we know that within the last million years incandescent pyroclast flows, some over 100 feet thick, have poured out over much of the central part of the North Island of New Zealand, travelling at over 60 m.p.h. and vaporizing all living matter over which they passed. Again, in the states of Washington, Oregon and Idaho in America, numerous floods of basalt lava are known to have been erupted several million years ago, some of which gathered into pools of molten lava 120 miles long by 50 miles wide and over 400 feet thick with an internal temperature in excess of 1000°C. Elsewhere, we know that in early historical times, around 1470 BC, the island of Santorini (Thera) in the eastern Mediterranean was partly destroyed in an extremely violent volcanic eruption, similar in many ways to that of Krakatoa, especially in that great tsunamis associated with it devastated surrounding shores and islands, possibly bringing the Minoan civilization to an end and giving rise to the legend of Atlantis. Archaeologists and historians have suggested that the death toll from the Santorini eruption could have been far greater than that of Krakatoa but, of course, there are now no direct records of the disaster that we can consult.

It is not surprising, then, that stories of volcanoes and volcanism are firmly implanted in legend and mythology. Homer, no doubt, was referring to Mount Etna, in Sicily, when he wrote of the one-eyed Polyphemus who hurled missiles at Ulysses. Even the otherwise prosaic Romans thought of volcanoes as the abode of Vulcan, who, forging thunderbolts for Jove, or the weapons of Cupid, lived in the subterranean dwellings of the Cyclops. Today, geologists regard volcanic eruptions rather differently, looking upon them as natural laboratories in which to further the ends of science. The contemporaries of Homer no doubt viewed volcanic eruptions with fear – the thunder and continuous roar of explosions, the ear-piercing detonations, the crash of falling missiles and the incessant rain of ash and dust, invariably accompanied by the relentless advance of sinuous black tongues of hot lava. Diodorus Siculus recorded that the Carthaginian army was turned back by an eruption of Mount Etna during their campaign against Syracuse in the year 396 BC. Little wonder that ancient man regarded volcanoes with superstition and awe, and thought of them as gods, or at least the abode of gods.

Before we can begin to understand the interrelationship of volcanoes with man, the nature of volcanic hazards and the possibilities of predicting volcanic eruptions, however, it is first necessary to know what volcanoes are, where they are found and why, and to review the products and styles of volcanic activity.

A volcano is generally thought of as a cone-shaped mountain, with fire and smoke spurting out of the summit. In general this is true, but with certain reservations; not all volcanoes have a symmetrical cone, not all breathe fire and the so-called smoke is actually cinders, ash and dust composed of finely pulverized rock and lava. A volcano is perhaps best described as an opening in the Earth from which gas or lava, generally both, are released. This means that a long fissure from which ash and lava are released is no less a volcano than the familiar cone-shaped mountain such as Mount Fuji in Japan and Mayon in the Philippines. The cone is simply formed by pyroclast material, thrown out of the vent by escaping gases, piling up around the opening to form an ever-growing mound of ash and lava; given sufficient time these cones can reach heights of over 13,000 feet. Volcanoes do not erupt all the time; they have rest periods. When there is a high probability that an existing volcano will one day burst into activity, then that volcano is said to be *live*; if it is actually erupting, then it is *active*; and if in a state of repose, it is said to be *dormant*. A great deal hinges, therefore, on whether or not a volcano can be determined to be live. Previously, a volcano was said to be active if it had erupted within historic times, irrespective of whether it was in eruption at the present. If it was not in eruption, then it was said to be dormant, while volcanoes which had not erupted within historic times were said to be *extinct*. Vesuvius was such a volcano – in Roman times it was thought to be extinct; yet in AD 79 it erupted violently. Again, Mont Pelée on the Caribbean island of Martinique was thought to be extinct, yet on 8 May 1902, as we have seen, that too erupted violently. Mount Lamington, in Papua New Guinea, was not even thought of as a volcano, let alone a killer volcano, in 1951, when 2942 people lost their lives. The town of Higaturu was destroyed and all its inhabitants were killed by the ash-laden blast of hot gases, and an area of dense rain forest, covering more than 90 square miles, was swept bare. Spasmodic activity continued at Mount Lamington until 1955.

Clearly 'historic times' is simply not a long enough period to determine whether or not a volcano is extinct. Some 100 miles north of Rome is the ancient Vulsini Volcano, the crater of which is occupied by picturesque Lake Bolsena. This, and other nearby volcanoes, laid down thick deposits of volcanic rock of simply enormous magnitude during prehistoric eruptions; the City of Rome stands on the deposits from one of these ancient holocausts. The time interval between these gigantic prehistoric eruptions is probably of the order of 5000 to 10,000 years, perhaps a little more, so there is a distinct possibility that volcanism may once again inundate the area around Rome at some time in the future. Recent work by volcanologists has shown that many volcanoes not known to have erupted within the past 2000 years are certainly not extinct. Many are undoubtedly still live, and must be regarded as potential killers. Only when it can be shown conclusively that there is little or no chance of an eruption taking place should a volcano be called extinct, or better still *dead*. Such volcanoes are normally very old and deeply dissected by erosion, sometimes deep enough to expose the rocks of the ancient magma chamber.

The kind of eruption of which a volcano is capable is governed by the type of rock it produces. Basically there are two main kinds of volcanic rock: *mafic* and *salic*. Mafic rocks are very dense and dark in colour; their lavas are runny, which allows gases to escape readily and hence prevents the build-up of dangerously high gas pressures. Chemically, these mafic rocks are rich in magnesium, iron, sodium and calcium and poor in silica, potassium and aluminium. Salic rocks, on the other hand, are poor in magnesium, iron and calcium but richer in silica, potassium and aluminium. Furthermore, salic magmas are thick and stiff like pitch and do not allow gases to escape readily. It is for this reason that salic volcanoes are able to build up dangerously high gas pressures, so that when they erupt they do so with explosive violence. Again by contrast with their mafic counterparts, salic rocks are light in colour and less dense: it is this difference in density that is the key to the understanding of volcanism.

Rocks at depth have conflicting physical properties. They are under intense pressure, yet are sufficiently plastic to permit slow circulatory movements under the influence of heat from deep-seated radioactive decay. Hot rocks expand, become less dense and, therefore, tend to rise slowly through the surrounding cooler rocks. Rocks are also sufficiently elastic to allow earth tremors to be transmitted through them, as we saw in Chapter 4, while at the same time being rigid enough to snap abruptly during severe faulting in an earthquake and, as magma, to break during volcanic eruptions to form jagged pumice fragments. Thus, igneous rocks at depth appear to have the impossible physical properties that are inherent in the strange compound called silicone putty: they are brittle, ductile, elastic and fluid.

We have seen that the Earth is composed of concentric layers of differing density and composition, and that radioactive heat is thought to be responsible for the creation of convection currents and mantle plumes within the Earth. We also know from earthquake studies that our planet has well-defined seismic zones and that the outermost skin of the Earth, which averages some 35 miles in thickness, is broken into six large plates and many smaller ones, rather like the shell of a cracked egg. New ocean floor is generated at the mid-oceanic ridges or spreading axes. These are often broken up into a stepped pattern by transform faults which take up the differences in spreading rates on the ocean floor. The increase in the surface area of the Earth by ocean-floor spreading is compensated for at island arc or continental margin trenches, where one plate is carried – subducted – beneath an adjacent plate, along an inclined seismically active plane known as a Benioff zone, to be destroyed and recirculated within the Earth. Submarine volcanism at the spreading ridges and in the great ocean deeps releases enormous volumes of lava onto the sea floor, but this activity is largely unseen and unrecorded. Volcanoes, as such, occur in three major and quite different environments: along the crests of constructive plate margins such as the mid-ocean ridges of the Atlantic Ocean and the East Pacific Rise; along the edges of destructive plate margins which border continental plates, such as the western cordillera of the South American continent and the western Pacific Island Arcs; and in the centre of both continental and oceanic plates, where they are located over more or less fixed hot spots or mantle plumes. The Hawaiian Islands, for example, lie in the centre of the Pacific Plate, the East African Rift Valley volcanoes are within the African Plate. Volcanoes, therefore, are not scattered haphazardly around the Earth: there are quite distinct patterns and these are governed by the distribution of what Earth scientists call plate boundaries.

Stretching right the way down the centre of the Atlantic Ocean is a chain of mountains and rift valleys. Very little is seen of this mid-ocean ridge, which appears at the surface only where volcanic activity associated with the ridge has been intense enough to build the ridge up above sea level. At the northern end of the ridge lies Jan Mayen Island, 930 miles north-east of Iceland, with, in its arctic desolation, the 7470-foot Beerenberg volcano which last erupted in 1970–1. Iceland, home of the famous geyser and land of ice and fire, stands astride the Mid-Atlantic Ridge and is actually still growing, for fresh molten rock or magma comes straight up from the Mantle to the Earth's surface here and is responsible for Iceland's 22 active volcanoes as well as the island's very existence. Each time magma fills a new fissure the rocks either side are pushed apart and, when this molten rock solidifies, the slight widening of the Crust is preserved at that point. Iceland is, therefore, being created along a broad north-east/south-west zone directly over the Mid-Atlantic Ridge: it is here that some of the newest rock on Earth is found.

Iceland has seen some of the largest volcanic eruptions in historic times. In 1783 an enormous flood of lava was released from a series of fissures 15 miles long running north-east/south-west across Laki Hill: in five months, nearly three cubic miles of new rock poured over 217 square miles of countryside. This, together with the fumes and heavy ash falls, nearly one-tenth cubic mile in volume, resulted in heavy losses to both livestock and population. The damage to crops, the death of a quarter of a million animals and the effect on the fish population caused a famine that killed some 10,000 people (about one-fifth of the population of Iceland at that time). More recently a new volcano, Eldjfell, appeared on the island of Heimaey just off the southern coast of Iceland which, during its five months of activity, partly buried the thriving fishing town of Vestmannaeyjar and nearly blocked the harbour entrance. The eruption began just after mid-night on 23 January 1973, when a fissure nearly a mile long ripped open from south to north like a giant zipper. Within minutes some 40 spatter cones, throwing out globs of molten rock, had grown along the glowing crack while lava poured down the hillside and into the sea. By good fortune the fishing fleet was in harbour and fuelled up ready to leave the following day. Within six hours the entire population of 5300 was safely evacuated without injury or loss of life. Heimaey is one of the Westmann group of islands of which the newest, Surtsey, appeared in 1963 and continued erupting until 1967.

Volcanic islands are next encountered 2800 miles south of Iceland, in the Azores, where submarine eruptions were reported in 1865 and 1884 from unidentified vents on the ocean floor. The Azores islands are all volcanic; eruptions took place from the perfect 7760-foot cone of Pico in 1963 and Capelhinos Volcano in 1957. The Canary Islands lie 930 miles south-east of the Azores and only 62 miles off the coast of the Spanish Sahara. Eruptions took place there as recently as 1902 and 1971. Further south again, 460 miles off the coast of Senegal, are the Cape Verde Islands where Fogo Volcano erupted in 1951. Although the Canary and Cape Verde Islands are not located directly over the Mid-Atlantic Ridge, they are included here because, firstly, they complete the picture of the distribution of volcanoes and, sec-ondly, they have affinities with volcanoes which lie directly over the ridge. From the Azores, the Mid-Atlantic Ridge curves westwards and then sweeps south-east, more or less parallel to the African coast, through the tiny islands of San Pedro and St Paulo to Ascension Island and St Helena, where the volcanic peak rises from a depth of

2½ miles to 2680 feet above the sea level. Halfway between the Brazilian coast and the Mid-Atlantic Ridge are another small group of volcanic islands, Trinidade and Martin Vaz. A few hundred miles north of the Antarctic drift ice, in the region of the roaring forties, lies Tristan da Cunha and its neighbours Nightingale and Inaccessible islands. An eruption on Tristan in 1962 led to the break up of the islands' community, when the population was prematurely evacuated to the United Kingdom; 310 miles further south is Gough Island, and 1500 miles south-east of there the lonely Bouvet Island, with its glacier-clad volcano, stands southern sentinel of the Mid-Atlantic Ridge.

The mid-ocean ridge turns east from here and passes south of the Cape of Good Hope into the southern Indian Ocean, where it becomes known as the Atlantic-Indian Rise. Passing through Prince Edward and Crozet Islands the Rise turns north 1240 miles east of Madagascar, near the Piton de la Fournaise volcano on Reunion Island, while the Mid-Indian Rise branches south-east. The northward continuation becomes the Carlsberg Ridge and passes west through the Gulf of Aden and up the Red Sea – a famous spreading axis which is slowly pushing Arabia north-east away from Africa. To the north of the Red Sea, and in part its continuation, the Dead Sea Rift (the lowest-lying land on earth) runs north into Syria and is there lost. To the south of the Red Sea, in Ethiopia, another branch follows the great rift valleys down into East Africa, where such famous volcanoes as Mounts Kenya, Kilimanjaro, Lengai, Longonot and Nyiragongo occur, with their own extension to Karthala Volcano on Comoro Island.

The Mid-Indian Rise runs south-east to the volcanic island of St Paul, then passes almost due east as the South-East Indian Rise, 1500 to 1860 miles south of Australia and New Zealand, to Buckle Island Volcano which last erupted in 1899. It turns northward suddenly 3100 miles from the southern tip of South America, and becomes the East Pacific Rise, passing through Easter Island and up to the Galapagos Islands where eruptions took place during 1970. From the Galapagos and north-west towards the Gulf of California we encounter the last two volcanoes associated with this mid-ocean ridge; Barcena Volcano on San Benedicto Island 370 miles off the western coast of Mexico, and the Tres Virgenes on the peninsula of Baja California. From here, the mid-ocean ridge runs beneath the North American continent causing complex volcanism, spreading movements and the earthquakes associated with the infamous San Andreas fault system.

The Earth is encircled, therefore, by major spreading axes on a system of mid-ocean ridges, and we have seen how closely these are associated with volcanism. We already know that new Crust cannot be continuously generated at these spreading axes without it being destroyed elsewhere, and that this destruction takes place where two plates collide at a destructive plate margin. These destructive plate margins, which practically surround the Pacific Ocean, have several features in common. On the landward side of the oceanic plate is a deep trench, while on the seaward side of the continental plate there stretches a 125–185 mile wide zone of volcanism. The margins of the plates here are also marked by an active seismic zone. Because some of the most active volcanoes of the world lie on these margins they are collectively referred to as the 'circum-Pacific Ring of Fire'.

The most southerly volcanoes of the world are found amidst the windswept wastes of the Antarctic continent where, in the desolation of the mountains and volcanoes of Marie Bird Land and on the peninsula of Graham Land, the most southerly tip of the vast Andean mountain chain is anchored. Deception Island and Lindenberg Island volcanoes are part of this chain, which runs north-east through the South Sandwich Islands and then curves through South Georgia to join South America at Tierra del Fuego. The Andes really begin here, although in these southerly latitudes there is little evidence of the high level mountain ranges which develop further north; but volcanoes start to appear among the multitude of islands and fjords which criss-cross the southern foothills. Monte Burney, the southernmost volcano of Chile, stands in lonely isolation many miles south of its nearest neighbour, Lautaro, while further north volcanoes occur in greater numbers. Passing northwards through Chile, Peru, Ecuador and Colombia we find some of the most active and dangerous volcanoes of South America and, indeed, of the world. In 1932, Descabazado Grande, in Chile, burst into eruption and its ash fall was carried by westerly winds as far as Rio de Janeiro, 1850 miles away, while large areas of Argentina east of the volcano were blanketed by white pumice ash. In 1963 the volcano Villarica, Chile erupted and melted large quantities of snow and ice to form destructive mudflows which cascaded down the mountainside causing considerable damage. Just inside the south Peruvian border, El Misti, a potential killer volcano, towers 11,480 feet above the nearby city of Arequipa which is built upon the thick deposits laid down by ancient, violently explosive eruptions of this same volcano. If such eruptions were to occur today, they would completely annihilate Arequipa City. In Ecuador, four large volcanoes occur which are consistently active: Reventador,

Cotopaxi, Tungurahua and Sangay have given enormous eruptions during historic times, producing ash columns as high as 15 miles and detonations which were heard as far away as 370 miles. In 1976, Sangay Volcano claimed the lives of several people on a climbing expedition. The Colombian volcanoes, Galeras and Purace, have behaved similarly to those of Ecuador. Although there are 59 catalogued active volcanoes on the South American continent, most of them fortunately lie in remote regions and therefore do not present any threat to mankind. If a census were taken of all the live volcanoes of these regions, including those which have not been classified as active for one reason or another, then the total could be as high as several hundred.

In Colombia the Andean volcano/mountain chain splits into two: one branch runs through Central America while the other branches north-east through Venezuela and into the many islands of the Lesser Antilles, before curving back through Haiti to join Central America in northern Nicaragua. There are 43 active volcanoes in Central America, some of which are very dangerous. In 1932, for example, glowing avalanches from Santa Maria, Guatemala, descended the valley of the River Tambor to produce boiling mudflows which caused tremendous damage and loss of life around the base of the volcano. Irazu Volcano, Costa Rica, produced such heavy ash-laden skies that thunderstorms gave torrential mudfalls, instead of rain. From Colombia the western branch runs up through Costa Rica, Nicaragua, El Salvador and Guatemala in an almost straight line of active volcanoes before reaching Mexico. Here, the chain is orientated nearly east–west from Veracruz in the east, to the Pacific coast west of Guadalajara; a few scattered volcanoes lie on islands just off the coast and there the chain dies out. The volcanoes of Central America and Mexico are very active and many are potentially dangerous. They include Izalco in Salvador, Fuego in Guatemala and Colima in Mexico. To these must now be added the volcanic complex discovered by one of the authors to the west of Guadalajara. The potential threat to this city, which is built on deposits from past big eruptions, was not realized until as recently as 1973, when the partly buried caldera of La Primavera volcano was located together with its fossil crater lake.

The volcanic chain appears again through the Cascade Ranges of the states of California, Oregon and Washington but with Mount Baker it dies out near Vancouver Island. Reappearing near Anchorage, the 'Ring of Fire' runs through the Aleutian Islands volcanoes to the peninsula of Kamchatka in eastern Siberia, a highly active region and home of the violently explosive Bezymianni volcano. South through the Kurile Islands to Japan we find such famous and destructive volcanoes as Fujiyama, Asama and Bandai San. Near Tokyo the chain divides once more, the eastern branch passing south through the Mariano Islands towards New Guinea while the western branch continues through the Ryukyu Islands, Taiwan, and the Philippines to Indonesia, the scene of the Krakatoa eruption. The western part of the chain is continued within the Indonesian Archipelago, but to the east it traverses New Guinea, the Solomon Islands and New Hebrides and finally passes down through the Kermadec-Tonga Archipelago to North Island, New Zealand. Isolated volcanoes such as Buckle Island in the South Pacific and Erebus Volcano in the Antarctic complete the 'Ring of Fire', although neither can be said to be associated with continental margin/island arc volcanism.

Many volcanoes isolated in the centre of crustal plates are now believed to lie above relatively fixed hot spots or mantle plumes (see Chapter 3). This means that as the plate moves slowly over such a hot spot a succession of volcanoes is produced, with the active one lying directly over the plume. In the Pacific a line of ancient submarine volcanoes, atolls and guyots stretches from the Hawaiian Islands to the WNW. What is interesting here is that their ages steadily increase away from Hawaii and this suggests that the Pacific plate, on its journey from the East Pacific Rise just off the west coast of South America to its destruction in the island arc trenches of Kurile/ Kamchatka and the Aleutians, passes over a relatively stationary hot spot whose roots lie deep in the Earth's Mantle beneath Hawaii, and that this is the cause of the volcanism there. Many other examples from the Pacific and elsewhere support this theory. Of the three volcanic settings we have described, continental margin/island arc volcanism is the most destructive and it is its volcanoes which, during historic times, have produced most of the world's most disastrous eruptions.

Volcanoes come in all shapes and sizes, and their activity is equally as variable, but more or less related to geological environment. Volcanoes on mid-oceanic ridges, for example, are quite different from those on continental margins or those found in continental rift valleys. There can even be wide variation in the type of volcanic activity within a single region. So, what governs this variation in activity and what is its possible social and economic significance? Different styles of volcanism are due to different factors, which may operate in isolation or in combination to influence the final result. Explosiveness is dependent on the viscosity of the molten rock; how fluid or pasty it is; and on the potential gas content. Magmas which release a lot of gas are generally responsible for very active and sometimes violent eruptions, especially if the rock is otherwise so viscous and pasty that the gas cannot readily escape. This means that the gas builds up to such enormous pressures that when it does escape it does so with explosive violence. The viscosity of the magma is governed by its temperature and chemical make-up; this in its turn is governed by the material from which the parental magma was formed in the first instance, and whether or not this magma has been contaminated, say, by melted-down fragments of the Crust during its long ascent to the surface. Volcanism at spreading ridges, mantle plumes and in the oceans generally is typified by the lava rock *basalt*. Basalt magmas are derived from partial melting of the Upper Mantle at depths between 30 and 100 miles below the Earth's surface. Volcanism in the island arcs and fold belts along destructive plate margins is typified by the additional presence of the lava rock *andesite* and by the even more salic rock *rhyolite*, which is more often present as pyroclastic deposits than as a lava. Andesite magmas are generated at similar depths to basalt magmas, by partial melting where a descending crustal plate in a subduction zone is consumed and reworked back into the Upper Mantle. Interaction between rising andesite magma and the deep continental Crust may be partially responsible for the production of the more salic magmas.

There are, however, many processes within the Earth that can operate in such a way as to change basalt or andesite magmas into a more salic melt. If a magma remains in a magma chamber for a long time, geologically speaking, then the chemical components of that mixture will slowly separate. In this differential separation of the heavier and lighter elements gravity plays an important part, and the more basic components of the magma will accumulate at the bottom of the reservoir, while the lighter, more salic ones will float to the top. When an eruption takes place from such a magma chamber, which has not released material for several hundreds or even thousands of years – and some volcanoes do have repose periods as long as this – then the first material to be released will be a relatively salic magma. It will have a high viscosity, contain a great deal of gas which has separated out as one of the lighter components, and be at a lower temperature than the magma at the bottom. As the temperature continues to fall near the roof of the magma chamber, the components begin to crystallize and the gases come out of solution, or exsolve; it is this accumulation of an unyielding magma highly charged with gas at the top of a magma reservoir which gives the potential for violently explosive eruptions.

The lower part of the magma reservoir, which contains the more fluid and more basic magma – the result of differentiation – will be relatively poor in dissolved gases. When this more fluid magma is released, its eruptive violence is less and, in the final stages of activity, lava may flow quietly from a previously violently explosive volcano as a degassed rock-melt. The normal course followed by any eruption is one of steady degassing of the magma.

The mildest form of volcanism is known as *hawaiian* type activity, after the central Pacific Islands where this form of activity frequently occurs. The lack of violence is due to the runny, low viscosity lava which allows the gases to escape readily. During hawaiian type eruptions great volumes of basalt lava can flow from fissures or small vents marked only by low spatter cones. However, even with these volcanoes, water can sometimes gain access to the magma conduits and the sudden conversion of this water into large volumes of steam causes exceptionally large steam blast explosions. Fortunately such explosions are rare. The fire fountains which commonly play over the most active vents are no more than gigantic firework displays and usually do not pose any great hazard. Quite often the degassed lava will quietly effuse from a bocca or opening at some distance from the main vent, the greater part of the available gas having been used to feed the pyroclast fountains. The main problem with highly mobile lavas such as these is that given a high discharge rate they pour over the sloping hillsides around the volcano to destroy farmland, sever communications and sometimes isolate villages, but sophisticated warning systems can normally ensure that loss of life is rare.

Hawaiian-style basalt fissure eruptions are those which produce the largest individual lava flows. Compound flow units more than 3000 feet thick are known to have occurred on some of the oceanic islands in the quite recent past. The Miocene Roza flow in eastern Washington, USA, covered an area of 20,000 square miles, and over most of this area it was at least 100–150 feet thick – its volume is in excess of 600 cubic miles of basalt! Thick basalt eruptions build up lava plateaux, many of which are to be found in the stratigraphic record of every continent. Examples include the early Tertiary basalt plateaux of East Greenland, the Faroe Islands and Antrim; the Miocene Columbia River Basalts in the USA; the late Cretaceous/early Tertiary Deccan Traps of India; the Cretaceous Paraná basalts of Brazil; and the Jurassic Karoo flood basalts of southern Africa. The areas flooded by these great outpourings of basalt lavas in the past were literally enormous: the Paraná basalts covered 500,000 square miles while the Karoo basalts probably covered as much as

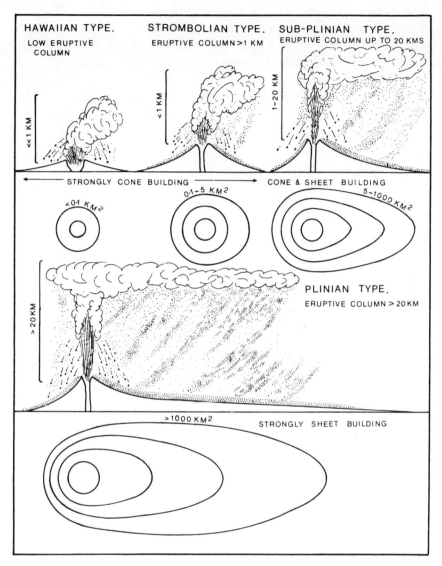

HAWAIIAN TYPE.
LOW ERUPTIVE COLUMN

STROMBOLIAN TYPE.
ERUPTIVE COLUMN >1 KM

SUB-PLINIAN TYPE.
ERUPTIVE COLUMN UP TO 20 KMS

<<1 KM

<1 KM

1-20 KM

←————— STRONGLY CONE BUILDING —————→ CONE & SHEET BUILDING

<0·1 KM²

0·1-5 KM²

5-1000 KM²

>20KM

PLINIAN TYPE.
ERUPTIVE COLUMN > 20KM

>1000 KM²

STRONGLY SHEET BUILDING

Fig. 23   Four types of volcanism illustrated here clearly demonstrate the relationship between viscosity, eruption size and area affected by air fall ash.

Viscosity is a measure of the internal friction of a liquid. It is low for basalts and high for granites and similar rock melts. Low viscosity magmas are thin and runny and this allows gases to form bubbles readily and easily escape without explosive activity. The hawaiian and strombolian types of activity are characteristic of low viscosity melts and produce low eruption plumes up to 1 km in height. The ejected debris is spread over a comparatively small area and the volcanoes are strongly cone building.

High viscosity magmas are thick and pasty, like pitch or toffee. This prevents the easy escape of gas bubbles. Because the gases are prevented from escaping they build up to high pressures, which culminate in violently explosive eruptions. Eruption plumes can easily exceed 20 km and the enormous volumes of debris are spread over an area as large as 1000 km².

Two types of material are thrown from the vent; those which are light are affected by lateral wind drift and pursue a non-ballistic trajectory, and those which are dense and heavy are unaffected by wind and follow a ballistic trajectory to land around the vent.

800,000 square miles. Volcanism on this gigantic scale will occur again in the future. When it does, reoccupation of the land between the rather infrequent flood eruptions will be attractive, especially to poor peasant farmers, because of the great fertility of the soils which quickly develop on basalt flows under optimum climatic conditions; such a reoccupation could only be recommended, however, if a proper scientific volcano surveillance is operative in the area.

The fire-fountaining which characterizes most forms of mild volcanism is commonly known as *strombolian* (or pyroclast – meaning hot-broken –) activity. Here, the escaping gases which are exploding in the vent of the volcano throw out large amounts of yellow-hot lava fragments and it is accumulation of this ejecta around the vent which builds up the familiar cone-shaped mound which is the popular conception of a volcano. In strombolian activity, which can vary from mild to violent, the fragments of escaping lava are charged with gas which, as it expands, distends the plastic rock into a porous cindery mass called scoria; the finer material consists of ash and dust which is frequently carried great distances by wind and may be the cause of agricultural losses far from the eruption site. An excessive gas content in bombs of lava thrown out by a volcano can cause these missiles to explode during flight, as in the 1948 eruption of Paricutin Volcano, Mexico.

Strombolian activity is named after the volcano Stromboli, a tiny island north of Sicily, known as the 'lighthouse of the Mediterranean' because it has been erupting continuously in this manner for centuries. Small strombolian 'cinder cone' volcanoes are scattered in their thousands over the volcanic fields of the world. The slopes of many large volcanoes, such as Mount Etna, in Sicily, are dotted with many of these so-called 'parasitic' cones.

Other forms of volcanic activity include *surtseyan* and *vulcanian*. The term surtseyan describes a form of violent steam blast pyroclast activity resulting from water repeatedly entering a lava vent that is at or just below the surface of the sea or lake. Enormous billowing white clouds of steam are produced, mixed with fine black ash and large lava bombs. This kind of eruption occurred during the early stages of the emergence of the new Icelandic island, Surtsey, from the Atlantic Ocean in 1963. Vulcanian activity is likewise named after an active volcano, the famous Vulcano in the Aeolian Islands, north of Sicily. Vulcanian eruptions are extremely explosive, and the debris thrown out is mostly finely fragmented pieces of semi-solidified vent-plugging lava mixed with fragments of the older lavas, subvolcanic intrusions, and pyroclast rocks of the cone. The amount of new lava involved in each explosion is usually quite small, thus vulcanian explosions, while often violent, are not the most violent form of volcanic activity. Large composite lava-ash volcanoes may be built by alternations of strombolian and vulcanian activity.

Further up the scale of eruptive violence there is the kind of activity reported at first hand by Pliny the Younger, whose eyewitness account of the eruption of Mount Vesuvius in AD 79, which destroyed the cities of Pompeii, Herculaneum and Stabiae, is the first scientific description of a volcanic eruption ever written. The eruption took place during the reign of Titus Caesar and is brilliantly described in two letters written by Pliny the Younger to his friend, the historian Cornelius Tacitus, telling of the circumstances of the death of his uncle, Pliny the Elder.

Sixteen years before the eruption, a severe earthquake had shaken the area around Pompeii and Herculaneum, and a number of buildings had been damaged. Further earthquakes took place around the volcano during the following decade, but little attention was paid to them (it was to be nearly 1800 years later before their significance was recognized by scientists). On 24 August AD 79 the increasing number of earth tremors finally culminated in the eruption of Mount Vesuvius, a volcano which at that time was believed to be extinct, although we now know that it had erupted some 1300 years earlier. In the space of a few hours the city of Pompeii lay buried beneath nearly 10 feet of volcanic ashes; nearby Stabiae suffered a similar fate. Most of the populace of Pompeii fled as soon as they realized the danger, although a few returned during brief lulls in the eruption to rescue their own valuables, or to loot the possessions of others. The populace left behind – and it is the fossilized remains of these unfortunates which are from time to time discovered during excavations at Pompeii (Plate 18) – probably comprises the elderly, the sick and infirm and those who returned for one reason or another. Tethered animals suffered particularly badly, as the remains of one dog testify (Plate 19).

At the time the eruption began, Pliny the Elder was stationed at Misenum, at the northern end of the bay of Naples. He was on active duty in command of the fleet; staying with him were his sister and his nephew, Pliny the Younger. The younger Pliny recounts:

> In the early afternoon of 24 August, my mother drew my uncle's attention to an unusually large cloud. My uncle was busy writing, after having bathed and lunched. Calling for his shoes, he climbed to a vantage point to get a better view of the eruption which, it transpired, was coming from a large mountain known as Vesuvius. The cloud was

like an umbrella pine, which rose on a great trunk with branches
splitting off. Sometimes it looked clean and at other times dirty.

It is quite obvious from this description that a great deal of steam was
admixed with the first volcanic ash that was being blasted out. Pliny
the Younger continues: 'My uncle perceived this phenomenon was
worthy of closer inspection and he therefore ordered a boat to be
made ready, telling me that I could accompany him if I wished.' The
younger Pliny, however, decided to continue with his studies and
therefore declined his uncle's offer. As his uncle was leaving the
building, he was given a message from Rectina, the wife of his friend
Bassus, whose home lay at the foot of Vesuvius. As she was in real
danger and could only escape by sea, she implored Pliny the Elder to
come to her rescue. The ships were launched about one hour after the
eruption had first been noticed and the original plan of scientific
inquiry was abandoned in favour of a rescue attempt to save, not only
Rectina, but many other unfortunate people who lived along this
beautiful stretch of coast. Pliny the Elder steered his ships straight for
the danger zone from which people were hurriedly leaving. 'My uncle
was completely without fear and gave instructions for each phase of
the eruption to be noted exactly as he saw them.'
     The time by now was between three and four in the afternoon and
hot pumice, ashes and blackened stones were falling in greater profusion as the ships drew near the coast, with the result that the shallow
water of the shore was choked with debris. Seeing that there was no
possibility of effecting a rescue here the elder Pliny put back to sea,
and sailed south to Stabiae in a full wind to meet Pomponianus who, it
seems, had already put his belongings on board ship, intending to sail
at the next favourable opportunity. Embracing the terrified Pomponianus, Pliny the Elder allayed the fears of his friend by staying the
night, and together they dined in the shadow of the angry mountain.
The volcano was belching sheets of flames at several points, which
Pliny suggested were bonfires or houses on fire, just to allay the fears
of his friends. Pliny the Younger goes on: 'My uncle then certainly
slept well that night, for his heavy breathing could be heard by all
passing his room. He was, after all, a corpulent man.'
     Meanwhile at Misenum there were violent earthquake shocks that
were also felt at Stabiae, where the ash level outside Pliny the Elder's
room had risen greatly.

My uncle was woken by his friends who feared they would all be buried
alive. They decided to chance their safety in the open as the buildings
were now shaking violently. Tying pillows to their heads as protection

against the falling stones, they ventured forth. The thickly falling ash and dust rendered the scene as black as any night and, by torchlight, my uncle and his friends made for the shore in order to ascertain the state of the sea. The sea was very rough indeed and escape by this route was not possible. Deciding to wait for better weather, a sheet was spread out on the beach for my uncle to lie on. During this time he asked repeatedly for drinks of water, then the approaching fire and stench of sulphur fumes soon compelled my uncle to rise. He stood up, leaning on two sticks and then suddenly collapsed.

It appears from the original text that at this point the elder Pliny was deserted for dead. When daylight returned on 26 August his body was found, unmarked and intact, for all the world looking like a sleeping old man.

It was on the 26th, as Pliny the Younger recounts, that severe earthquakes at Misenum led to near-panic among the population there, as they fought and jostled to escape from the danger of their collapsing homes. 'Tidal waves' left fish stranded while the volcano erupted with ever-increasing fury, pouring out enormous black clouds rent with lightning. One such black cloud was seen to pour over the land like a great flood of water, blotting out the sea at Capri and Misenum. This final event was, no doubt, one of the most terrible of all manifestations – a pyroclast flow. While Pompeii was being steadily buried on the lower slopes of Vesuvius, vast quantities of pumice and ash were accumulating on the upper slopes. Torrential rain, which so often accompanies large eruptions such as this, saturated these ashes, triggered their collapse and caused them to avalanche down the side of the mountain. These ash/mud slurries, which gathered up more and more debris as they swept down the side of the volcano, eventually poured into Herculaneum and buried the town. Herculaneum has now been partly excavated, and the fact that no human remains have been found there suggests that the population had time to evacuate before their town was completely inundated by the mudflows.

Because Pliny was the first to describe the type of eruption which occurred at Vesuvius in AD 79, such outbursts have become known as *plinian* type eruptions. They are characterized by a continuous or steadily reverberating gas blast over a comparatively short period of time, quite often a matter of hours, or two or three days at the most. During this period of extreme eruptive violence, enormous volumes of volcanic ash and pumice fragments are ejected in the uprising *eruption column*, where they are entrained. These eventually fall around the vent, or downwind from it, to mantle the countryside thickly, like an enormous snowstorm. As in the case of Pompeii in AD 79, and nearly so with Vestmannaeyjar in 1973, towns can virtually disappear overnight beneath a thick blanket of ash.

The area covered is generally known as the dispersal area: an elongated dispersal fan is produced when an eruption takes place in a wind and a circular dispersal area is produced when the eruption occurs under quiet meteorological conditions. The largest and heaviest fragments fall closest to the vent and the finer and lighter particles settle farther away, their distance from the vent depending on how quickly they can fall through the atmosphere – their terminal fall velocity. Careful study of the degree of sorting shown by the deposits which result from *pyroclast fall activity* can tell the volcanologist how particular volcanoes have behaved in the past, and their past activity is the surest guide to how they are likely to behave in the future. Obviously, the shape of the dispersal fan can tell us whether a wind was blowing during the eruption and also its direction. Providing certain assumptions are made about the height of the eruption column, then it is possible to calculate the relative wind speed. The shapes of the *isopleths* – lines of equal value, denoting size in this case – for maximum pumice and maximum lithic (dense solid rock) fragments can tell us whether the vent or blast was angled or vertical. From such distributions we also can get an indication of the shape of the eruption plume, whether it was narrow or widely diverging. Measurement of the total thicknesses of ash over the entire dispersal area permits us to calculate the total volume of material erupted and compare eruption magnitudes – based on ash volume – of different volcanoes. This is a factor of prime importance when investigating possibilities of prediction and hazard mitigation.

All kinds of secrets, therefore, are locked up in deposits from old volcanic eruptions. Other studies of volcanic ash deposits can tell us if it was raining during the eruption, for when it is, spherical mud pellets called accretionary lapilli form as the raindrops fall and are tossed about in the turbulent eruption cloud. These *accretionary lapilli*, when found in ancient volcanic deposits, are essentially fossil raindrops; and undisputed proof of heavy rainstorms. Larger accretionary lapilli are most certainly formed during extremely violent thunderstorms, when strongly rising air currents saturated with volcanic ash and dust encourage the formation of mud hailstones. Accretionary lapilli due to such phenomena, which were produced during terrible eruptions in prehistoric times, have been found on several large Italian volcanoes such as Vulsini, Sabatini and Laziale: these same volcanoes which, we have suggested, may not be dead and should, therefore, be treated as potential killers. Other features in volcanic deposits tell of violent lightning displays within the swirling murk of the eruption cloud/thunderstorms. The uprushing column of steam, gas and debris and the movement of tiny particles of rock and water swirling around in the churning and billowing fountain vomited by the Earth generate vast amounts of static electricity. It was this static electricity which caused the St Elmo's Fire on the mast heads and rigging of ships caught near the eruption of Krakatoa in 1883. The fierce momentary heat which is produced when this electrical energy is dissipated during a lightning strike from the ground produces glass melt tubes coated with pumice lumps, known as *fulgurites*, in volcanic deposits.

Plinian eruptions do not always stop after the initial high velocity vertical release of large volumes of ash and pumice; there is often a change in eruption style which leads to an even more deadly type of eruption: the type which was first brought to the attention of mankind by the twin disasters of 1902 in the West Indies. Albeit on a much smaller, though nonetheless deadly, scale than the many similar eruptions that must have taken place in prehistoric times, detailed scientific study of the form and products of the eruptions of Soufrière and Mont Pelée in that year has enabled geologists to understand one of the world's most terrifying potential earthshock phenomena – *pyroclast flow volcanism*.

Towards the end of the plinian phase of an eruption, the explosion level in a volcanic vent may fall as magma is used up, so the rising gases must expend more energy in expanding to surface pressure along an ever-lengthening pipe. This creates a gun effect, and although the amount of gas may be actually less than that during the opening stages of the eruption, it can achieve more by virtue of the work it does during expansion. With continued fall in gas pressure, however, the eruption column diminishes in height and the gas blast is often insufficient to carry all the material completely clear of the volcano summit. The result of this is that the outer, heavy, basal part of the eruption column may slow down and finally suffer a total collapse. The downward falling material rapidly settles around the vent of the cone as an annulus which, because of its rapid rate of accumulation, is extremely unstable. Under the influence of gravity this annulus fails and cascades down the sides of the volcano as an avalanche of incandescent ash, dust and lava fragments. These pyroclast flows (as opposed to the pyroclast falls discussed above) behave as fluids, because the particles of rock within them are buoyed up by an envelope of hot gas surrounding each tiny fragment. Sometimes called nuées ardentes or glowing avalanches, pyroclast flows can move at speeds of over 100 miles per hour and, because of their high degree of fluidity, they may follow river valleys, ravines and similar topographical depressions. On a smaller scale these killer clouds can also be produced by a kind of vigorous activity within the vent which mills the rock – as if in a giant grinder – without actually throwing it clear. Eventually the entire fluidized mass may pour over the lip of the crater rather like a saucepan of milk boiling over.

Another kind of nuée ardente is produced by high gas pressure rupturing the base of a lava spine protruding from a volcano or by the explosive collapse of a lava dome. Both of these latter mechanisms can also produce large masses of fluidized, highly fragmented rock particles at high temperatures buoyed along by equally high temperature gases; it was such a killer cloud which annihilated St Pierre in 1902. Nuées ardentes give little or no warning, travel at hurricane or atomic blast speeds and are hot enough to kill instantly by asphyxiation or cadaveric spasm and carbonize the corpse by destructive distillation. They are like an incandescent rock aerosol and without doubt are one of the most terrible of all volcanic manifestations. Volcanologists gain further knowledge from every eruption that is scientifically observed: they have obtained much new data on explosions from atomic tests and from the detonation of large amounts of TNT. Base surges driven by hurricane force winds are a feature of atomic explosions, and we now know that they are a common feature of many large volcanic or other explosions too. When water gains access to the open vents of volcanoes erupting in the sea, or similar aqueous environments, for example, large steam blast explosions frequently occur in which a horizontal surge component may form as well as the upward directed blast. Base surges take the form of a very rapidly expanding low level annular cloud: on land they are sometimes called ground surges to distinguish them from those which occur in aqueous environments. Deposits of base and ground surges are like those of pyroclast flows in many ways and often precede them.

Throughout geological time, innumerable pyroclast flows of enormous magnitude have been erupted, but, so far as we know, none have been witnessed by trained observers. In 1912, Mount Katmai in Alaska erupted and released a large pyroclast flow which travelled 14 miles down a valley, burying the countryside under some 100 feet of hot ash. Steam escaping from underlying vegetation and groundwater rose through the ash for many years in a multitude of steam jets or fumaroles, hence the name 'Valley of Ten Thousand Smokes'. By comparison with some prehistoric flows and those of the more distant geological past, however, the Katmai event was quite small. Today we know that there are many live volcanoes still capable of producing pyroclast flows which could wipe out and completely inundate an area half the size of the State of California in a few short hours of incandescent horror!

Repeated pyroclast flows build an extensive plateau as they bury the underlying land surface. One of the largest pyroclast flow plateaux known to science covered much of Nevada and the adjacent parts of Utah some 20–30 million years ago. Within this great volcanic field, one giant pyroclast flow has been identified that covers 7000 square miles and contains 250 cubic miles of rock! The Bishop Tuff, exposed in the north end of the Owens Valley in eastern California, is a single 200-feet thick pyroclast flow that covered around 400 square miles of the western USA some 700,000 years ago. Enormous, geologically young, pyroclast flow fields are known in the North Island of New Zealand; in Arizona; New Mexico; Colorado; Wyoming; Sumatra; Japan; and many other places. Small centres of pyroclast flow activity have been found throughout the world's volcanic areas. In the more distant geological past, many gigantic volcanic floods are known to have been produced by variants of the pyroclast flow mechanism, one example of which is the 189 million year old Karoo Rhyolite formation of south-eastern Africa. Pyroclast flows of a different type are also known to occur beneath the sea.

In 1875, the Icelandic volcano of Askja erupted with such vigour that much of the air fall pumice was sufficiently hot on landing to weld together to form a hard compact rock. Similar eruptions have occurred in the past on Pantelleria island in the Mediterranean, and we know that in the geological past many volcanoes throughout the world have behaved in a similar way, to produce what we call *welded air fall ashes*. Burial of a large city by volcanic ash is a terrible fate for any community, but burial by falling globs of molten rock which are hot enough to weld themselves together is a horror of mind-boggling dimensions. The *pièce de résistance* of catastrophic volcanic phenomena appears to be equally divided between large pyroclast flows and large volume welded air fall ash, both of which have yet to be experienced by civilized Man.

Constant accompaniments to many forms of primary volcanic activity are landslips and mudflows. Volcanic mudflows and torrent deposits may be either hot or cold and the waters involved may be highly acidic; they are called *lahars* and can be killers, as we have seen in the descriptions of the 1953 disaster in New Zealand, when 151 people travelling on the Wellington/Auckland express were killed, and at the Rivière Blanche in Martinique, in 1902, when 23 workers in the Guerin Distillery lost their lives. In fact, lahars have killed many thousands in historical times alone. They are particularly prevalent around andesite cones with high level crater lakes, and are one of the common products of plinian volcanism. Pyroclast flows often degenerate into lahars as they penetrate long distances down river valleys.

Months, and sometimes years, before a volcanic eruption takes place there are disturbances which may pass unnoticed by even the most experienced observers. The reason for this is that the disturbances take the form of minute seismic shocks – tiny earthquakes – and these can often be detected only by sensitive scientific instruments. Larger shocks may be felt but pass unheeded; all too often mild earthquake shocks are attributed to some mundane cause such as the passing of a heavy lorry. These faint shocks are caused by magma moving upwards towards the surface of the Earth, prising the rocks apart in a series of cracks. As the magma approaches the surface it rises faster and so the number of shocks increases. Shortly before an eruption takes place the number of seismic shocks may rise until the ground is continually quivering, on a microscopic scale. Animals are able to detect these disturbances and this may explain why many leave the area around volcanoes shortly before eruptions occur. The dogs on the Warner Ranch, five miles east of Pahoe, in Hawaii ran around digging holes and snuffling excitedly in them three days before an eruption began a few hundred feet from the ranch. The microseismic shocks had obviously attracted their attention.

When a new volcano is to be born, a fissure or series of fissures opens where the pressure is greatest, or a line of weakness exists, and after the soil and rock covering have been blown away by escaping gases, small globs of lava appear. These lava droplets may coalesce to form an actual lava flow, like those which develop from the fire fountains in Hawaii, or they may be thrown high into the air and thus cool before they land around the vent to build up a cone. When large established *polygenic* volcanoes erupt, the warning seismic shocks still occur, but the actual outburst is usually a much more vigorous affair than in the case of a *monogenic* volcano, which erupts only once.

The story of the activity of any particular volcano is one of steadily falling gas pressure; often the activity appears to increase as the eruption progresses but this may be due to the gases expending more energy in expanding to surface pressure as the depth of explosion increases (a point mentioned earlier). Volcanoes may also show vigorous fire-fountaining at the summit while degassed lava flows quietly out at the base of the volcano, an abrupt separation of the two phases. Later in the life of an eruption, magma may be so degassed that a pyroclast flow may form, or a lava spine or dome may grow in the vent; this may also be followed by pyroclast flows.

After ash and pumice have stopped falling and lava has ceased to flow the volcano enters a fumarolic stage when only residual heat and gases are evolved. This stage, which may last for many years, is dependent on the heat reserves of the lava and ash. Generally thick deposits produce the most lasting fumarole activity, although a slow seepage of heat to the surface and an active water table may allow fumaroles to vent steam and gas for many years without the presence of ash or lava fields at the surface. Areas of very hot ground may overlie a magma chamber roof and provide a useful and continuing source of geothermal power. Larderello in Italy, Rotorua in New Zealand, and the geysers in California and Iceland will provide cheap sources of central heating and electricity for many years. Similar sources of energy are now being sought in many countries, with varying degrees of success: since the 1973 eruption on Heimaey, for instance, heat exchangers installed on the thick lava flow promise over a decade of cheap central heating.

We have seen that a pattern of world volcano distribution can be plotted, associated with the mid-oceanic ridges, continental margin/island arc subduction zones and mantle plumes, but there are no established patterns of volcanism with regard to the timing of eruptions. Occasionally, when several volcanoes within a discrete zone become active, this activity may be due to abnormally high crustal movement on a subduction zone. Areas of the world which are affected by highly destructive earthquakes may also be associated with such movement. On 18 January 1902 an earthquake caused extensive damage near Quezaltenango and along the Pacific coast of Guatemala. On 17 April, Quezaltenango was destroyed and 1000 people killed. On 25 April, Mont Pelée, in Martinique, started erupting. On 7 May, Soufrière on St Vincent Island in the Caribbean erupted violently and on the next day, Mont Pelée, only a few miles to the north, vented its fury on St Pierre. On 10 May, Izalco volcano in Salvador erupted and this was followed by Masaya, Nicaragua on 15 July; Concepcion in Nicaragua, and Santa Maria and Pacaya, both in Guatemala, all erupted in 1902. It appears significant that seven major volcanoes from the Caribbean Plate area all erupted, some violently, within the space of that one year. But the volcanological record of activity is too incomplete as yet to permit cycles of such events to become an established fact. However, the science of *tephro-chronology* allows the volcanologist to examine deposits from ancient eruptions, which occurred before historic records began, and they can tell us what happened in prehistoric times and, given dateable carbon samples from the volcanic ash, when it happened. As a result of recent advances in volcanology we are now certain that many volcanoes exhibit a crude periodicity. Perhaps future research in closely linked zones will establish patterns which are related to both time and periods of accelerated plate movement.

Not all volcanoes are killers. There are some 516 catalogued active volcanoes throughout the world, but of these only 89 are known to have caused casualties during historic times. In the casualty league, Japan tops the list with 14 disastrous eruptions, Java is second with 12 and the Philippines and Central America equal third with 6 each. Tambora Volcano in the Sunda Islands caused the greatest certain loss of life during an eruption: in 1815 an extremely violent outburst destroyed the top of the volcano and reputedly discharged 24 cubic miles of pumice ash. During the eruption between 56,000 and 80,000 people are reported to have been killed. However, in the 1669 eruption of Etna, it is estimated by some authorities that 100,000 people may have lost their lives in the destruction of the city of Catania. In modern times, it is Krakatoa and Mont Pelée that are better known as the killer volcanoes. It was the St Pierre disaster that first aroused serious concern about the risk from future eruptions which could give rise to nuées ardentes.

As a result of our recent research work in western Mexico, evidence has been revealed which tells us, without any doubt whatever, that the city of Guadalajara is quite literally sitting on a time-bomb of atomic or cataclysmic proportions. Painstaking attention to detail here has made it possible to draw up volcanic hazard maps which define areas that would be at risk during the next large eruption, while statistical calculations on the data obtained give an indication of the likely size and characteristics of the next outburst. It is therefore possible to give a fairly accurate scenario of the events that may take place during the next violent outburst. Unless well-thought-out contingency plans are prepared in advance, the authorities will be helpless and tens of thousands will die.

Some time before the actual eruption, perhaps a few months or even several years, there will be a number of small earthquakes which may be felt far below La Primavera Volcano in Guadalajara City. These quakes will be caused by molten rock ascending from depth and slowly forcing its way through and rupturing the solid rock as it intrudes the lower confines of the volcano to heat up the long dormant magma chamber. This influx of heat will provide the necessary impetus to set the whole volcanic apparatus operating again, the magma chamber will be reloaded and, due to the increase in volume, the volcano will expand. This expansion can be measured at the surface by sensitive scientific instruments, which will indicate where swelling of the volcano is greatest and therefore where the eruption is going to take place. Unfortunately the probability is that no such measurements will be made; the swelling will go unnoticed and the

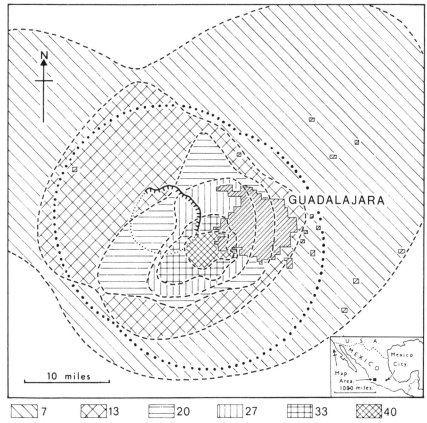

Fig. 24   Volcanic risk map: Guadalajara, Mexico, showing the percentage probability of burial beneath one metre or more of ash during the next big eruption. Also shown (as thickly dotted line) is the extent of areas at risk from heavy 10 cm missiles. NE scarp face of caldera shown as heavy ornamentation.

Guadalajara City is virtually on top of this potentially dangerous volcano and there is a high probability of extensive damage or burial during future activity. Pyroclast flows from this volcano have affected most of the area shown on the map. A future event would practically wipe out all life in an unevacuated city.

earthquakes may be attributed to passing heavy vehicles or some other such cause. Even the increase in the temperature of the hot springs which abound amid the lava hills of the central part of La Primavera complex may fail to attract attention, or at best be put down to yet another natural curiosity. In the absence of a local seismograph, only the larger earthquakes may be noticed as the total number of 'quakes per day progressively increases. By the time the first gas and steam jets break surface it is unlikely that anything will have been done to warn the population of the impending danger; in fact the chances are that nothing will be done. 'It can't happen to us' will be the cry, or 'It's just another Paricutin', and the preliminary activity will, no doubt, be treated as a lucrative tourist attraction. Meanwhile, the eruption will have been inexorably marching towards its dreadful climax. The blast of hot gases will have been enlarging the original fissure and the activity will have been steadily growing more violent, with the appearance of great ash-laden clouds above the vent. To those living near the volcano the eruption will now be assuming proportions vastly greater than those of 'yet another Paricutin'. Loss of communication as roads progressively become blocked by ash and as telephone wires break could quickly lead to panic in the population. Ash will be falling thickly around the volcano and downwind of it, while enormous thunderstorms, caused by convection in the rapidly rising eruption cloud, will produce terrifying lightning displays and localized torrential rain. This rain will mix with the thickly falling ash to produce ugly grey mudflows which will pour across the slopes of La Primavera Volcano to engulf nearby homesteads and make any serious attempt at rescue even more hopeless.

Once a volcano such as this has started to erupt violently the course of events usually proceeds very quickly and the climax may occur within 48 hours. The scenario painted so far is one of confusion and chaos, but there is far worse to come, for the volcano has only just embarked upon a series of events that could culminate in the total destruction of a large city and most of its inhabitants. As the volcano develops its final onslaught, destructive explosions may occur as large volumes of steam are produced from ground water circulating beneath the surface. Torrential rain and mudflows would fill the caldera and produce perhaps a temporary lake, which could be instrumental in generating the destructive horizontal blasts that are so often associated with volcanoes which erupt below water level; the massive pumice blocks from these explosions would form a huge raft of floating rock, just as they have done in the past. The pumice is so distended by gas bubbles that it is less dense than the water and therefore able to float. These horizontal blasts, or base surges, consisting of pulverized rock at high temperature held in suspension by equally high temperature gases, will sear across the countryside at hurricane-like speeds, burning nearly everything in their path and demolishing homes and buildings with terrifying ease.

The eruption column will now be ascending ever higher, sending pumice and ash as high as 20 miles or more; the roar of the continuous gas blast will be heard for many scores of miles and irregular explosions as far away as several hundred miles. The thickly falling ash will impair the functions of both men and machines and attempts to leave the stricken area by car will be doomed to failure, for the ash will be too thick to drive through, the air inlets of the engines will quickly block up with fine volcanic dust and visibility will be down to only a few feet in places. Any attempt to leave vehicles and proceed on foot will be fraught with immediate danger; the choking dust could easily cause death by asphyxiation in many, while others may be struck by falling rock from the eruption cloud, against which they will almost certainly not be protected. In the few cases where cars are able to be driven there will be no chance of escape, for the monumental traffic jams that will result due to panic and fear will very quickly bring any evacuation attempt to a halt. Only those fortunate enough to live on the outskirts of the city, away from the volcano, will have any reasonable chance of escaping. The hopeless situation will drive hundreds into churches and cathedrals where they will invoke their god, or gods, for deliverance from this terrible destruction.

There is now only one final ritual to be enacted by the volcano. As the gas pressure falls the weakening blast will be unable to hurl the vast volumes of pumice high into the atmosphere and the eruption cloud base will collapse around the vent to produce base surges; alternatively the volcano will mill and churn the rock within its throat. This finely churned-up rock will then be released suddenly as a thick cloud of hot dense gas, highly charged with incandescent rock dust and magma droplets, and its sudden expansion as it clears the throat of the volcano will cause it to rush outwards from its source with the speed of a thermonuclear blast. It is, in fact, similar to the base surge but larger by an order of several magnitudes. Sometimes there is a deceptive lull in activity before the deadly pyroclast flow is released. A plug of lava may rise up the throat of the volcano to quite effectively block it. To any survivors this may herald the end of the eruption, but the calm is deceptive and only hides the fact that gas pressure is steadily building up again beneath the lava plug. Perhaps the plug will hold in the contents of the volcano and the eruption will end here; after all that is what happened after the last eruption of Tequila volcano, only a few miles to the north. If the plug fails, as they so often do, then the pyroclast flow will be released with tremendous violence. The clouds of destruction will sweep across the countryside up to many scores of miles from the volcano, destroying everything in their paths. Instant death by cadaveric spasm or asphyxiation will overtake many and others will die later from the terrible burns that they will receive. Guadalajara City will be flattened as if by a thermo-nuclear bomb blast and scores of outlying villages will, quite literally, be wiped out. Even after all this the region will not be safe until the last lava plug, or dome, which normally follows this type of activity, has finally ceased to grow.

This is just an isolated instance of what will certainly happen to one of the world's largest volcanoes in the future. Throughout the world man is steadily encroaching upon the lower slopes of volcanoes everywhere, for after all, why not? The rich volcanic soil is an open invitation to cultivate and farm. Violently destructive eruptions have happened many times in the past on such volcanoes, however, and the odds are that they will most certainly happen again in the future. Farmers who persist in cultivating the slopes of these potential killer volcanoes may, quite literally, be digging their own graves.

6   Frozen Death or Flooding

And now there came both mist and snow,
And it grew wondrous cold:
And ice, mast high, came floating by,
As green as emerald.

And through the drifts the snowy clifts,
Did send a dismal sheen:
No shapes of men nor beast we ken –
The ice was all between.

The ice was here, the ice was there,
The ice was all around:
It crack'd and growl'd, and roar'd and howl'd,
Like noises in a swound!

> Samuel Taylor Coleridge, *The Rime of the Ancient Mariner*

'At the sound of the great waters marching over the land,
the very salt of the earth trembles in its dreams . . .'

> Saint-John Perse

From the geological record, two certain predictions can be made: the first is that the world sea level will oscillate against the land by several hundred feet in the relatively near future. Oscillations of sea level by various amounts up to 250 feet have occurred during the last 30,000 years, that is, since man began leaving a record of his activities in cave paintings and other similar graffiti – and these changes in world sea level are additional to any local subsidence or elevation of the land. Secondly, it is equally certain that the polar ice cap will once more expand and bring glacial conditions and land ice back to much of northern North America and northern Europe. The last time this happened on a massive scale was only 18,000 years ago, while the retreat of the northern ice from the last small continental advance occurred as recently as 10,000 years ago. Thus the future prospect for many of the presently prosperous industrially advanced countries is either frozen death or flooding. The only uncertainty is which will come first.

We have seen in previous chapters that the present distribution of land and sea is ephemeral on anything other than a very short-term time scale. The geological record is full of great transgressions of the oceans across the continents, followed at some later date by relative uplift and the re-emergence of these shallow sea floors as new land. The dance of the continents, particularly the growth and decline of mantle plumes, sea-floor spreading ridges and the development and closure of oceanic deeps at destructive plate margins, is continually changing the relative volumes of the ocean basins. This, in its turn, causes the ancient stable cores of the continents to stand alternately high and low relative to world sea level. The whole process is immensely complex. It is possible for some parts of a continent to be inundated while other parts are being uplifted slowly, by mantle plume or up-doming or in an isostatic response to unloading, or sharply, in the fold belts caused by plate collision. Again, long-term subsidence may be typical of certain areas of a continent, and this subsidence will be accentuated by the increasing weight of sediment that tends to accumulate in any kind of persistent downwarp or basin. The North Sea, between Britain and the rest of Europe, is such an area. Added to this never-ending *geotectonic* (earth movements) variation in local, regional and worldwide relative sea levels, there is the periodic extraction of massive volumes of ocean water to form great ice caps at the North and South Poles of our planet.

Ice is not always present at the poles. When both polar regions are free from permanent ice, the Earth experiences a warmer, more equable climate. Nevertheless, we know from the record of the rocks that there have been a series of great ice ages in the past, during each of which large ice caps have grown at the poles and glaciers have appeared on all high mountains, even those in the tropics. Not much is known of extremely ancient ice ages, but we have considerable data on four major glacial periods that occurred respectively in the middle Precambrian (around 2250 million years ago), the late Precambrian (580–1000 million years ago, with complex maxima around 950 ± 50, 777 ± 40 and 616 ± 30 million years ago); the Permo-Carboniferous (between 325 and 250 million years ago) and the Quaternary (the last 2 million years).

What causes ice ages? One cause is clearly the drift of a continent or group of continents across a pole. One of the reasons why there is now a great ice cap on the Antarctic continent is because this continent is at present, once again, passing over the South Pole during its interminable dance around the globe. Again, while there is not a continent at the North Pole today, the Arctic Ocean is surrounded on all sides by continental land and has few exits: even the greatest of these, that into the Atlantic, has a shallow water sill in places. It is very likely that when, as the result of continental drift, there is no land near either pole, the formation of great ice caps is inhibited; for, if both polar regions are at the centre of unrestricted oceans, the sea ice that grows every winter will be dissipated during each succeeding summer, and no permanent ice cap will accumulate. At the time of the Permo-Carboniferous major ice age, it was the very old nucleus of Pangea, the so-called Gondwanaland group of continents, that was drifting across the South Pole. Massive *tillites* (ancient ice-deposited boulder clays) of this age are found in South America, South Africa, India, Antarctica and Australia. In fact, we can trace the history of these continents as they drifted across the South Pole throughout Palaeozoic times. In the Ordovician, North Africa was at the pole and tillites were deposited by ice in the Sahara. By Lower Devonian times, central South America was at the pole and by the late Carboniferous, eastern Antarctica had moved into this position. In Permian times, the Gondwanaland group of continents were moving rapidly westward and the South Polar ice cap was centred progressively further towards Australia. The size of the ice cap did not remain stable, however, and in late Carboniferous to early Permian times it was immensely bigger than it had been in the Ordovician. The presence of continental land at the poles thus enables polar ice caps to grow, and continents drifting across the poles will experience ice age conditions locally, but this combination of circumstances is not sufficient, in itself, to explain the advent of the four major ice ages we listed above. During the late Precambrian complex of major ice ages, for example, glacial conditions were exceptionally intense and widespread over the whole globe, and ice cap conditions were experienced on almost all continents, even those far from the poles. During this same glacial period Antarctica was one of the few areas free from ice! Again, as we have just noted, the maximum glaciation of the Permo-Carboniferous major ice age occurred during the late Carboniferous and early Permian, while Gondwanaland was moving across the South Polar region. Before the beginning of Triassic times, however, even though the edge of the supercontinent was still close to the pole, glaciation ceased abruptly. Continental drift, therefore, including its probable influence on world climatic zonation, is but one factor in the generation of ice caps.

The other main factor controlling the growth of ice caps is a change in the heat energy balance within the hydrosphere and atmosphere of our planet. This energy balance change will be revealed as a change in climate and weather patterns. The solid Earth itself is, of course, a source of surface heat which is continuously lost into space via the sea and air. The heat supplied from the solid Earth has fallen steadily over its lifetime, and the clustering of mineral forming events in the Crust around certain times does suggest a long-term cyclic periodicity in this source of heat energy. Nevertheless, measurements show that because the amount of energy arriving from the Sun every day is almost 6000 times greater than that reaching the surface from below, the principal factor influencing the energy balance of the hydrosphere and atmosphere – and thus world climate – is change in the average annual radiant energy received by our planet from the Sun, rather than variations in the Earth's internal heat supply.

For too long scientists have tended to regard the output of energy by the Sun as a virtually unchanging constant, but this is not so. Careful observations made over many years, combined with intelligent interpretation of historical records, have shown that the Sun's energy output does indeed fluctuate, on both a long-term and short-term basis. We now know that the presence or absence of the great storms that are visible in the surface layers of the Sun, called 'sun spots', makes a considerable difference to its energy output. Sun spot activity appears to rise and fall in a combination of 11-year and longer-term cycles. When sun spots are completely absent, the radiant energy being received by the Earth probably falls off dramatically. The last time sun spots were completely absent was between AD 1650 and 1700. This period of reduced solar energy reception is called the Maunder Minimum, and the result of it was an increase in the size of the polar ice caps and a general deterioration in world climate. In Britain, the River Thames froze over every winter in London and milk came home frozen in the pail. The build up of ice lagged behind the minimum on the solar energy cycle and its maximum advance did not occur until around 1750, after which it began to retreat once more. The increase in sea ice as a result of this late seventeenth- to early eighteenth-century 'little ice age' may be why so many attempts between 1600 and 1800 to find the North-West Passage around Canada were defeated by excessively Arctic conditions. Little ice ages comparable to this event appear to recur on an approximately 2000-year cycle, previous examples being known around 0–500 BC and 2000 BC.

Minor fluctuations in solar energy levels cause even smaller 'little ice ages' and short periods of climatic deterioration. Examples of these include several minor glacial advances known in historical times: around 1350 BC, 750 BC, in the Middle Ages and more recently. On the Arctic island of Jan Mayen, for example, the glaciers show evidence of having retreated from their AD 1750 advance in an oscillating fashion, with readvances occurring around 1850/70, 1900/10 and 1950/55, each of which was, however, lesser than its predecessor. The minor fluctuations of solar energy indicated by little ice ages of all scales and the intervening episodes of exceptional warmth have a profound influence on the world's weather patterns, and even small shifts in the weather patterns can have dramatic effects in some areas of the globe, leading to drought and/or famine earthshock. The exceptionally hot dry summers of AD 764 and 783 caused drought, famine and town fires in Britain, France and Germany, and many people died as a result of the hot weather. The thirteenth century is known to have been characterized by generally warmer conditions and high sea levels throughout the Northern Hemisphere. In Colorado, USA, a twelve-year drought beginning in 1273 decimated the population of Mesa Verde, driving many Indians from their ancestral hunting grounds.

We do not know what causes the long-term fluctuations in the reception of solar energy that determine whether or not we have major ice ages on Earth. Various suggestions have been made: most of these involve the periodic passage of the solar system through clouds of interstellar gas or dust, some of which is swept up by the Sun. While the fall of this interstellar material onto the surface of the Sun could certainly result in an increase in the output of solar energy being radiated towards the Earth, it is also possible that reception of radiant energy by the Earth might be seriously obstructed by the very same infall. That the Sun may meet with considerable variation in the distribution of interstellar debris, dust and gas during its long voyage around the galaxy cannot be denied. The long-term periodicity of warm intervals and major ice ages may be directly related to galactic rotation, or to the movement of the solar system through the meteorite, dust and gas rich spiral arms of our galaxy, or it may be quite random: with our present knowledge we can do no more than speculate on the possibilities.

For mankind, the relevant question is: given that at infrequent intervals there are periods of time during which the overall reception of solar energy by the Earth is significantly different from the norm, are there events on the Earth itself which can cause what might have been otherwise no more than a tendency towards more glacial conditions to escalate into an episode of major ice advance? Clearly the atmosphere, which is a filter that can pass or reject solar radiation, is of prime importance in this respect. When it is highly polluted with reflecting dust or vapours very little of the Sun's heat will be transmitted. For example, it has been suggested that the immediate effect of a sudden large rise in the level of solar energy being received might be vastly to increase the cloud cover throughout the world. A virtually continuous thick cloud cover formed over the whole globe in this way would create an efficient reflecting layer and, while it lasted, would prevent much of the solar heat reaching the Earth's surface.

Another possible explanation for the occurrence of ice ages has been sought in the periodic polarity reversals of the Earth's magnetic field. During these changes the shielding effect of the *magnetosphere*, the Earth's magnetic envelope, is dramatically reduced, allowing the Earth to be exposed to a much higher level of cosmic radiation. This would lead to increased ionization at the level of the *tropopause* (the top of the near surface realm of air circulation, clouds and weather), and so promote the formation of a blanketing layer of highly reflective cirrus clouds by nucleation.

Recent research has also suggested a correlation between some glacial advances and large volcanic eruptions. After explosive volcanic eruptions a great deal of highly reflective dust enters the atmosphere and may remain suspended for many years, causing a global decline in temperature due to its shielding effect. During such declines, springs and summers are made even colder as a result of the additional winter snowfall increasing the reflectivity of the Earth's surface by amounts that may be as large as a factor of four. This combination of events is believed to be instrumental in producing a surfeit of ice and, consequently, enhanced glacial activity. Some support for this thesis can be obtained from an examination of historical data from South America. In 1835, there was a great plinian eruption from Coseguina Volcano, Nicaragua, and from several other volcanoes in the period 1843–6. These outbursts were followed by very cold springs and cold summers: the wine harvests were late, crops failed and there were food shortages in 1838–9. Glacial advances followed during the period 1840–55. Again, major episodes of volcanic activity in the southern Andes 5450 to 4700 and 2850 to 2150 years ago were followed by glacial advances during the periods 5260 to 4590 and 2940 to 2100 years ago respectively. It is clear that if solar heat can be virtually cut off at ground level as a result of a reflecting layer developing high in the atmosphere, the effect on the weather and on snow accumulation below would be cumulative and accelerating. 'Instant glaciation' might be the result. Even after the reflecting layer was dissipated, it is probable that once a critical amount of snow and ice had accumulated in high latitudes, an ice age begun in this way might be almost self-perpetuating.

Man emerged on Earth just before the advent of one of the world's major ice ages: the first reappearance of excessively glacial conditions since the Permo-Carboniferous glaciers melted and Gondwanaland finally became free from ice some 250 million years ago. Over the last two million years our planet has been in the icy grip of the Quaternary Ice Age, during the course of which great ice caps have grown at both poles, and on several occasions – perhaps as many as 20 times – have expanded far beyond their present confines. In the Northern Hemisphere, ice sheets have covered most of northern North America and northern Europe on at least four occasions. At its maximum advance, the North Polar ice stretched from the Ukraine to the mid-west of America. In Britain it reached as far south as London. Most Earth scientists are convinced that the current 'great ice age' is not over and that we are living in what is no more than an interglacial period between major ice advances. (*Interglacial* periods are time intervals during which the climate temporarily improves and the ice retreats.) The deposits of the Quaternary Ice Age record many such interglacial periods. In the longest of them, the climate was probably warmer and more equable than at present. Each interglacial (and the shorter *interstadials*) most probably had its own series of 'little ice ages' and intervening warmer periods. For modern man, the crucial question is, therefore, when will the present interglacial period end? It began around 10,000 years ago with a great melting and disappearance of ice sheets. A climatic optimum occurred around 3500–3000 BC, since which time there have been at least three sharp deteriorations of the weather and many smaller fluctuations. The next deterioration in our weather could well herald the return of full glacial conditions! Alternatively, we could still be progressing, albeit in a hesitating manner, towards the mid point of the current interglacial. If this is the case, we can look forward to a generally warmer worldwide climate and further ice retreat towards the poles in the reasonably near future. Whichever of these possibilities actually ensues, the consequences will be equally disastrous: we cannot ignore the fact that we are living in a period of unstable climatic equilibrium within one of the major 'ice ages' of the Earth's history and that a quite small change in the levels of solar energy being received at ground level could bring back the ice to much of the Northern Hemisphere or, alternatively, melt much of the ice at present locked up in the polar ice caps!

According to old Norse legend, a summer which remains snow-bound from the previous winter and in which the snows fail to melt is not a summer at all and will herald the end of the world. In 1783, the gigantic fissure eruption at Laki in Iceland released vast quantities of dust which heavily polluted the Earth's atmosphere for long after the eruption ended. The pollution was so bad that when Benjamin Franklin wrote from Paris that year he complained that a constant fog, which the sun could not penetrate, hung over Europe and North America and that the snows and ice of that year remained largely unmelted. As we saw in Chapter 5, the atmospheric pollution resulting from the 1883 eruption of Krakatoa caused a dramatic fall in the level of radiant energy being received from the Sun at ground level – for many months between 10 and 20 per cent less energy was received at places as far away from Krakatoa as Europe. Laki must have had an even greater adverse effect, but no comparable figures are available. In early 1977 the United States was hit by a series of blizzards of unprecedented ferocity, which caused such severe conditions that a state of national emergency was declared. Cities and their communications were totally paralysed, people were unable to reach their place of work and everyone was advised to stay indoors for, outside, 50–60 mile per hour blizzards were blowing with temperatures as low as minus 50 degrees Centigrade. Townships were cut off, thousands of cars buried, sometimes with their occupants, and hundreds of people lost their lives. If something had occurred to prevent the melting of this exceptionally heavy snowfall, like a large number of big volcanic eruptions (and the history of the Earth contains evidence of many eruptions considerably larger and more violent than either Laki or Krakatoa), then perhaps we would be already well on our way into the next ice age.

Ice ages happen very suddenly. The old idea that glaciers extend south from their northern domains to slowly but relentlessly overrun the Northern Hemisphere must now give way, in the light of new findings, to the more catastrophic hypothesis that ice ages start very suddenly and that snow and ice accumulate right here where we stand. Glaciers do not come creeping down from the north; what actually happens is that scattered snow and icefields simply coalesce to produce an 'instant' glacial area! Repeated winter snowfalls of 18 inches or more which do not melt completely during the following summer would have a devastating effect, and lay enormous areas of

Europe and North America a desolate waste under an ever-increasing blanket of snow.

As the thickness of snow increases, its extra weight compresses it into a granular matrix called *névé* or *firn*, then, with further increase in thickness and pressure this névé is transformed into ice. When a large mass of ice exceeds a certain thickness then the pressure in the centre causes it to flow outwards; this is why the polar ice caps maintain a more or less constant thickness. Although ice appears to flow as a plastic mass the actual mechanism is much more complex: the intense pressure of the thick layer of overlying ice causes the boundaries of each of the ice crystals of which the glacier ice is principally composed to melt and form a tiny film of water surrounding each crystal. This water acts as a lubricant and allows the crystals to slide over one another easily. Then, when the pressure is removed, the water freezes back to ice. By these means huge masses of glacier ice can move over vast distances.

Ever since the first gases appeared from the cooling Earth and began to circulate in response to the transfer of heat from one region of the globe to another an atmosphere has been in operation. Atmosphere is the 'fluid' medium which largely controls the heat balance of our planet and is responsible for the daily, and long-term, weather patterns; by its response to the incoming solar radiation it governs whether or not our lands will be buried by ice, turned into scorching deserts, or drowned by rising seas and oceans as fresh water is released by melting polar ice. So how was our life-supporting atmosphere created?

After the creation of the Earth some 4600 million years ago, its early history was one of saturation bombardment by planetisimals. A combination of various processes may have caused partial or total melting of the hot proto-Earth. In any event, as it cooled from this state, it contracted and some of the lighter components were sweated out as large volumes of water vapour, carbon dioxide, ammonia, methane and various other gases. This sweating out was most probably achieved by a form of volcanic activity. Further cooling permitted certain components of this gaseous mixture to condense and form the first primitive clouds. As the cooling of this primeval atmosphere continued, the water vapour condensed, first to clouds and then to water, which fell as rain on the hot primitive land surface. Most of this rain, or rather solution of gases, was vaporized as soon as it touched the hot surface, but eventually the land surface cooled sufficiently for streams to form and then seas to develop in land depressions. Even today, gases and water vapour are released every time there is a volcanic eruption, and rain is always a solution of gases, which is why plants grow better when irrigated with rain water than with tap water. Only a part of the water and gas released by volcanic eruptions is, however, primeval, the remainder is recirculated groundwater and water that was trapped in ancient sediments within the Earth's Crust.

After the formation of the early atmosphere, the lighter gases such as hydrogen and helium escaped into space – or were driven off by the solar wind – leaving behind a mixture of water vapour, methane, ammonia and formaldehyde. Under continual bombardment by ultraviolet light these molecules broke down into hydrogen, nitrogen, carbon dioxide and water. Hydrogen produced in this way again escaped into space and thus the Earth's primitive atmosphere gradually became enriched with nitrogen and carbon dioxide. The first macromolecules of replicating proteinoids might have formed from this primitive atmosphere, as a result of electrical discharges during violent storms, solutions of the gases in rain falling onto hot lava, or acoustic shock waves during highly explosive volcanic eruptions passing through clouds. Alternatively, as is suggested by Sir Fred Hoyle and N. C. Wickramasinghe, life may have arrived on Earth from interstellar space, carried to our planet on a meteorite. Whatever the truth of the matter, once the initial creation of life on Earth had occurred, primitive plants were able to develop in the hot springs and pools which are normally associated with waning volcanic activity and to colonize the growing oceans. From that time on, oxygen has been continuously released into the atmosphere as a result of the reduction of surplus carbon dioxide in a biological reaction within living plant material until, today, oxygen and nitrogen form over 98 per cent of the gases of the air; this figure does not, of course, include water vapour, which varies in amount from place to place depending on atmospheric temperature and water availability.

Climate is the result of a combination of the circulation of water about the oceans of the Earth and the incessant wanderings of this gaseous mixture we call our atmosphere. The heat which drives this gigantic climatic engine is supplied as solar radiation by the Sun. The amount of solar heat falling on the Earth and its envelope at any one point is dependent upon several variable factors, the most important being the Earth's attitude to the Sun (i.e.: its seasonal position and periodic wobble) and the distance of the Earth from its source of heat and life energy.

The axis of the Earth tilts by 22° to 24° every 40,000 years relative to its elliptical orbit round the Sun. This affects the range of the seasons while the eccentricity of the orbit causes up to 30 per cent seasonal change in the intensity of solar radiation received at any locality: at present the northern hemisphere is experiencing a 6¾ per cent seasonal change. The complex interrelationship of these factors resulted in a particularly large seasonal change some 10,000 years ago, at the end of the last ice advance in the Northern Hemisphere. As the Earth spins around the Sun, following its elliptical path, it tends to 'wobble' about its axis, in the same way as a spinning top will wobble – this motion is caused by a number of different factors, one being the gravitational pull of the Sun and the Moon acting on a slightly non-spherical Earth and it is this wobble which causes the apparent wheeling pattern of the star constellations seen from the Earth. It follows a regular pattern, known as precession, repeated over a period of some 25,800 years. Another, smaller, wobble is caused by the orbit of the Moon around the Earth, and the shape of the Earth, with its bulge around the equator, results in a regular, 'nodding' pattern repeated over a period of about 18 years.

All these variations are mainly of interest to the astronomer, but there is another short-term effect which may be significant in our consideration of the Earth as a working whole. This is the Chandler Wobble, named after the American amateur scientist who first described it in 1891. The main feature of the Chandler Wobble is a regular small change in the direction of the axis of rotation of the Earth, occurring during a period of 14 months, after which the effects die away over a period of 10–30 years. The initial stimulus which causes the wobble has been the subject of controversy ever since it was first discovered; the main causes that have been suggested at one time or another in the past include large-scale atmospheric movements – the gravitational piling up of masses of air in certain areas on the surface of the globe – causing variations in the tilt of the axis; various interactions of the Earth's Core and Mantle have also been suggested as a possible cause; and, most significantly, earthquakes. The physical size of major earthquakes has been measured very accurately with precision instruments and this new knowledge, together with the most recent information about the direction of crustal movements caused by plate tectonic motions, makes it possible to estimate the cumulative effects of large earthquakes on polar motion – that is, the slight change in direction of the axis of rotation which constitutes the Chandler Wobble.

A study of the major earthquakes that have occurred so far during this century has shown a close correlation between the global pattern of seismicity and the variations that have been observed in the Chandler Wobble over the same period. The effect of any individual earthquake, however large, is too small to be detected; it is the cumulative effect of a large number of earthquakes which appears to bring about the variation. Another factor, which may be of equal importance, is the aseismic movement along the boundaries of plates deep in the Crust – this movement is not directly associated with earthquakes and is, therefore, very difficult to measure. It certainly seems that major earthquakes could account for a large part of the mechanism which excites the wobble from time to time, and also for the long-term variations in its amplitude; the remainder could be due to a variety of factors, such as the large-scale movements of the Earth's atmosphere. The pattern of the Chandler Wobble over the last 70 years has shown a large amplitude in 1910 which decreased until 1920 and then remained fairly small until 1940. Then it began to increase again and reached a maximum in 1954 since when there has been an irregular decline in its amplitude. It now appears that large earthquakes are probably as effective in damping down the wobble as they are in exciting it; the decrease in the wobble from 1914 to 1929 coincides with several large earthquakes which took place during this period.

Most of the solar heat which falls on the outer layers of our atmosphere is lost in reflection, scattering and absorption during its attempt to reach the surface of the Earth. Of the total solar heat being received, only 43 to 48 per cent actually reaches the Earth's surface to be absorbed as short-wave radiation by the land surface or oceans, where it may be transmitted to considerable depths. About half of this solar energy reaches the ground directly through the atmosphere; the remainder arrives indirectly at the surface via atmospheric scattering and diffusion through cloud and water vapour layers. Some 16 to 18 per cent of the incident radiation is lost through absorption in the stratosphere and troposphere, while up to a third of the solar radiation reaching the Earth's outer layers is reflected back into space. These values naturally apply only to solar radiation received during the hours of daylight. Most of the heat energy actually absorbed by the atmosphere comes from a long-wave reradiation of solar energy from the surface of the Earth. This energy radiation from the ground or water surface may be reflected and reradiated back to Earth during particularly cloudy conditions, when as much as 80 per cent is returned. This is why clear nights are much colder than cloudy ones, even in the summer or in the desert, for, although the atmosphere does limit the immediate loss of radiant heat, when the insulating layer of cloud is absent much more heat is lost into space.

The percentage of solar radiation absorbed at any one place is dependent upon the reflectivity of the absorbing material (its *albedo*). Therefore, while black soils and similarly dark sediments with a low albedo will reflect only 7 to 20 per cent of the radiation they receive, ice and snow, which have a high albedo, will reflect as much as 90 per cent. The high latitudes of the polar regions receive less solar heat because the radiation must follow a long oblique path through the atmosphere and cloud layers to the surface and thereby subject the incoming heat energy flux to a greater loss by absorption. How much of the incoming heat is actually passed by the atmosphere will depend on changes in atmospheric moisture content and cloud cover, and also on the carbon dioxide concentration and volcanic dust content. An increase in the carbon dioxide content can act as a filter which allows the solar radiation to reach the Earth's surface more readily while preventing some of the heat from Earth being lost to space. Any adverse combination of these factors could seriously affect our climate and initiate a major climatic change.

The unequal heating of the various parts of the Earth's surface causes great inequalities in the density of the atmosphere near to and parallel to the Earth's surface. Masses of air at different densities will tend to circulate about until they reach an equilibrium position or until their temperature changes sufficiently to establish a different density relationship. Less dense warm air rises and spreads out over the relatively more dense cold air, which sinks and spreads out under the rising warm air. The permanently greater heating of the equatorial regions is seriously affected by the presence of the polar ice caps. The equatorial zones, which receive over twice as much heat per annum as do the poles, release some of this heat to drive a circulation of the atmosphere from the equator to the poles. This gives a degree of permanency to the climatic zones parallel to the latitudinal circles. The atmosphere is like a giant heat engine, doing work against the ground friction of geographic irregularities such as mountain chains, while transferring heat from the surface of the warm equatorial regions to the cold sink of the upper atmosphere, where more heat is lost through radiation to outer space than is gained through absorption of incoming radiation from the Sun and outgoing radiation from the Earth. Were the source of solar radiation suddenly cut off the whole atmospheric circulation engine would very quickly come to a standstill.

In the equatorial tropics the less dense warm air rises from the heat equator, which does not coincide with the geographical equator, up a narrow corridor known as the Intertropical Convergence Zone to produce a pressure gradient between the warm equatorial zone and the cold polar zones. In very general terms, this means that the less dense warm air extends polewards at high altitudes, while its place is taken at low altitudes by relatively denser cold air streaming from the polar regions towards the equatorial zone. As the Earth rotates it does so relative to the atmosphere which surrounds it and this produces a movement of the atmosphere over the Earth's surface which we call wind. This rotation also deflects air streams to the west in equatorial regions and results in a clockwise (right hand) movement around high pressure centres in the northern hemisphere and a counter clockwise (left hand) movement around high pressure centres in the southern hemisphere. The converse applies for low pressure centres.

The mainstream of the atmospheric circulation engine is the predominance of the upper westerly jetstream circulation, strongest in the middle latitudes and forming stable circumpolar whirls from a height of two-thirds of a mile to high into the stratosphere. These whirls dominate the weather patterns just outside the tropical zones between latitudes 30° and 60°; occasionally their influence will extend nearly to the equator. Ultimately they are responsible for all the weather that we experience here on Earth.

Hurricanes, or tropical cyclones, probably originate from waves in the tropical easterly air-flow, particularly when the atmosphere is in an unstable condition to the height of the westerly subtropical jet

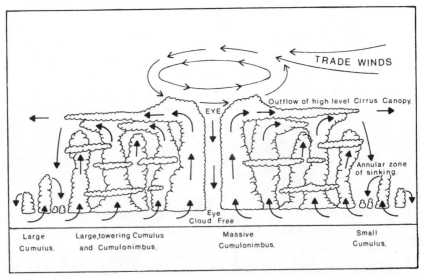

Fig. 25 Cross-section of a hurricane showing the directions of wind movement and the development of various cloud types. At ground level the wind speed increases towards the centre of the whirling vortex until velocities in excess of 100 miles per hour are reached. The strong updraft around the core causes immense towers of massive cumulonimbus thunderheads to form, while there is a steady downward flow of air inside the relatively calm, cloud-free core.

stream. Their energy source is the ocean water which, when evaporated, converts latent heat into kinetic energy and real heat which spirals rapidly upwards around a calm central column, or eye, of descending cooler stratospheric air. The winds spiralling round the eye, clockwise in the northern hemisphere and anticlockwise in the southern, transfer the heat collected from the ocean surface to the cold sink of the upper atmosphere and in doing so create rapidly rotating storm systems 50–500 miles across. At the top of the eye of a hurricane the upward airflow spirals out in the opposite direction to which it ascended and the water vapour condenses to produce an outward moving canopy of cirrus clouds at the level of the stratosphere. It can, therefore, be seen that hurricanes require heat and evaporation from the oceans to sustain them; when they pass over land they very soon die. However, the frictional drag of the land surface causes the winds to spiral inwards more rapidly and rotate more quickly (just as a skater increases his speed of rotation when he holds his arms close to his body); as a result of this, hurricanes are most dangerous immediately after they come ashore.

In 1969, hurricane Hanna smashed into Galveston, Texas and in three days caused terrible damage and personal loss. On the second day of the storm, the sea, lashed to relentless fury by the hurricane winds, topped the sea wall and flooded the town centre. The sea front was absolute chaos as high seas repeatedly crashed onto the shore, sending up 100-foot high waves. The beach and sea were littered with wreckage, partly from the demolished pier and partly from boats and shore-line huts. Only after Hanna had spent her energy was it possible to take stock of the situation. Whole rows of houses had been smashed like so many wooden boxes and there was damage as far as the eye could see, dazed members of families were wandering about, having lost nearly everything and now carrying all they possessed. People of the stricken area congregated at the soup kitchens, and mobile canteens were set up to feed children for whom it was sometimes the first meal in several days. For the army it provided at least an exercise in putting their amphibious tanks and vehicles to good use, for without their assistance the delivery of the all-important drugs to medical centres and food to the hurricane relief distribution centres would have constituted an additional problem. During the onslaught of hurricane Hanna some strange things happened: a 2 by 12 inch plank was driven right through a tree, without breaking, and the flooding caused poisonous snakes to seek refuge in the town suburbs. Water Moccasins and Rattlesnakes became a danger as they took refuge on verandahs.

In the Northern Hemisphere the great mountain chains of North America, northern Europe and Asia form geographical barriers which, by virtue of the friction they cause, produce a braking effect which can deflect the high altitude westerly air streams. In the Southern Hemisphere, however, where over 80 per cent of the area is ocean, there is little friction and the region has no effect other than to ensure the stability of the southerly circumpolar whirl. The overall pattern of circulation between the equatorial zone and the polar zones takes place through a series of interconnecting atmospheric cells, each of which has its own well-defined pattern of circulation and which can vary in extent from season to season. Indeed, it is the location, size and vigour of these cells which determine climate and the onset or recession of glacial periods, although these characteristics are, in their turn, governed by more complex processes in our turbulent atmosphere. Between latitudes 30° North and South positioned roughly either side of the equator, are the two Hadley Cells. They meet at the heat equator to form the Intertropical Convergence Zone (ICZ), where great upsurges of warm moist air take place as a result of surface heating of the air in the north-east and south-east trade winds. These upsurging masses of moist warm air form towering cumulonimbus thunderclouds as they enter the upper, cooler, zones of the atmosphere. When they are between 9 and 12 miles high (49,000 and 65,000 feet), the air streams of the Hadley Cells turn away from each other and travel diagonally polewards, their upper surface being at the level of the Tropical Tropopause, all the time losing heat. At around 30° latitude the high level air flow of the Hadley Cells encounters the high level air flow of the mid-latitudinal Ferrel Cells, streaming towards the equatorial zones on its return journey across the temperate zone away from the polar regions. This confluence reacts to produce the westerly sub-tropical jet streams, which blow at hurricane force around a height of 8½ miles (46,000 feet), and a combined sinking of the cooled air towards the Earth's surface. The sinking air flow produces a stable zone of high pressure, with little or no rain, known as the sub-tropical high: at low levels, the outflowing air travels equatorwards in the Hadley Cells as the trade winds and polewards in the Ferrel Cell as the westerlies. Around 60° latitude, the Ferrel Cells abut against the Polar Cells in a complex airflow of disturbed Westerlies and Westerly Polar Front Jet Streams along a slanting polar front above the zone of extratropical cyclones and anticyclones. The ceiling of the Ferrel Cell, the sub-tropical (mid-latitudinal) tropopause, decreases in height from 9 miles to 6 miles (49,000 to 33,000 feet) towards the polar front. The

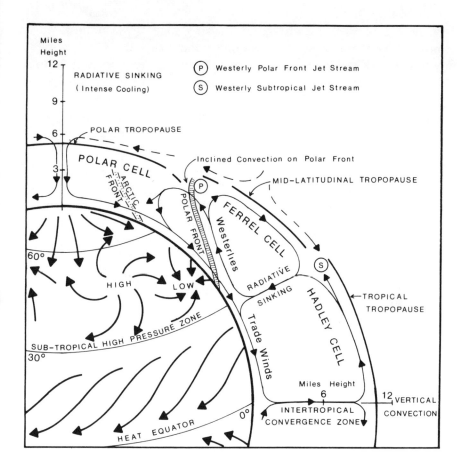

Fig. 26 The Earth's climatic engine: here, in a diagrammatic section of part of the Northern Hemisphere, the airflow pathways of the world clearly demonstrate how the climatic zones of our planet are interrelated. In equatorial regions, moisture-laden air rises rapidly along the Intertropical Convergence Zone into the cooler zones of the upper atmosphere. The cooling effect in this zone of vertical convection causes towering masses of cumulonimbus thunder-clouds to develop; this accounts for the heavy rains experienced in tropical regions. The circulation of this, now dry, air of the Hadley Cell just below the tropical tropopause meets a similar air stream from the Ferrel Cell and the combined, cooled, airstreams undergo radiative sinking to produce the sub-tropical high pressure zone at the Earth's surface. The air streams separate here; one flows south as the trade winds, while the other flows north as the westerlies. The corriolis effect of the rotating Earth deflects these airstreams in the direction opposite to that of the Earth's rotation.

The polar front is caused by the cold air of the Polar Cell forming a wedge beneath the warmer northern margin of the Ferrel Cell. The warmer air rises over the cold wedge in a zone of disturbed westerlies marked by an alternating pattern of high and low pressure anticyclones and cyclones, the position of which is closely determined by the position of the westerly polar front jet stream. Complex airflow patterns which are associated with the polar front are responsible for the vagaries in the weather of the Northern Hemisphere. The arctic front (and corresponding antarctic front) demark the junction between the intensely cooled polar airflow and the disturbed southerly wedge of the Polar Cell. (In the case of the southern hemisphere the compass directions are, of course, reversed.)

Hadley and Ferrel Cells together form a wedge of warmer air which moves polewards along the inclined polar front above the corresponding wedge of cold air of the Polar Cell, which is sinking equatorwards. Two further fronts, the Arctic, and the corresponding Antarctic front, form a partial division within each Polar Cell and separate the disturbed westerlies from the circulating polar airflow which spirals out, after intense cooling due to radiative sinking, from some 5½ miles (29,000 feet) above the poles. Great eddies of air at the Earth's surface are related to the jet streams, and make the characteristic weather of the stormy temperate zones in which many of the industrially advanced countries are situated. Changes in the routes followed by the jet streams can make great differences to the climate of these countries.

The oceans are the second most important medium in determining climate, for water can retain its absorbed heat for longer than the atmosphere and will influence the temperature of the air above the ocean, which in turn will alter the water temperature elsewhere. The net effect of the surface winds blowing over oceans is to set in motion the surface layers of water, and establish a pattern of circulation which may be augmented by vertical movement of waters of varying density and temperature. The picture of oceanic circulation is affected by the rotation of the Earth in much the same way as our atmosphere, for, as our planet spins it deflects the waters of the oceans in the direction opposite to its rotation and causes a hydrostatic head to build up in the region of the sub-tropical high pressure zone, near latitude 30°. A condition exists whereby the outflow of this head of water is in equilibrium with the deflection caused by the Earth's rotation. With the spin effect thus nullified, the major circulatory motion of the oceans generally mirrors that of the prevailing winds.

The oceans absorb vast quantities of solar heat and, although much of this heat is dispersed, there is still sufficient left to have important effects on our climate. For example: the Gulf Stream carries warm water from the Greater Antilles and Florida coasts across the Atlantic Ocean to play on the westerly coasts of Ireland and Scotland and sustain a flora which is worthy of more southerly latitudes. These ocean currents may act to produce ridges of thermal highs and lows which are superimposed on the Earth's general atmospheric pattern. The effects of these ocean currents would be greatest when the thermal difference between the poles and the equator was high and least when the thermal difference was low. The circulatory flow of the oceans is augmented by the ocean floor flow of cold polar melt water towards warmer regions, to replace the northward flow of warmer surface waters.

World climate is governed by the amount of heat which the Earth receives from the Sun and its stability is maintained by the precarious balance between the atmospheric filter, the terrestrial reflector and the absorptive powers of each. So, we have the eternal chicken and egg argument. The weather pattern, with its clouds and water vapour distribution, governs how much solar radiation reaches the Earth at any particular point, while it is the solar radiation which creates the distribution of clouds and water vapour. Therefore, the purer our atmosphere is, the less water vapour and hence fewer clouds there are, and the greater is the input of solar radiation and its consequent heating effect but, because of the differing degrees of reflectivity of the Earth's surface, the *greater* will be the temperature gradient and hence pressure gradient between the polar and equatorial regions. Conversely, the dirtier our atmosphere, whether in terms of natural components such as water vapour and cloud, wind blown and volcanic dust, or in terms of artificially introduced components such as exhaust gases from the internal combustion engine, contents of aerosol containers and the varied obnoxious aerial effluents of modern industry, the *less* will be the intensity of atmospheric circulation. On the other hand we do not want our atmosphere to be too pure for, in theory, it could let in too much heat and melt the polar ice caps. If incoming solar radiation increased to the point that the polar ice caps melted completely, the atmospheric temperature and pressure gradients would decrease until they hardly existed, and the air flow system would very quickly slow down.

What then really exerts the greatest control over the future of our climate? A reduction in the incoming solar radiation at ground and sea level would decrease the thermal and pressure gradients between the polar and equatorial regions. The consequence of this would be a shift towards the equator by the Polar Fronts and the compression of the Ferrel and Hadley Cells. Certainly, when the Earth's surface is colder, the intensification in circulation would increase the snowfall in the higher latitudes of the Northern Hemisphere to the extent where more snow would accumulate during the winter than could thaw during the following summer; this would be particularly evident in Scandinavia, Greenland and northern Canada, and over high ground such as in the mountains of the Alps, Himalayas and Rockies. However, the tilt of the Earth on its axis would probably emphasize the deficit of solar energy received at one particular pole relative to that at the opposite pole and encourage an unequal growth of new ice.

The ice in polar regions would not melt so readily as before because of the colder weather and the high albedo of new snow and ice.

In contradistinction, increased heating of the polar regions could seriously affect their stability, eventually reaching a stage where snow of one winter was insufficient to last until the following winter and so preserve the ice caps. Of the two hemispheres, it is the Antarctic, a flat expanse of polar ice surrounded by a great ocean uninterrupted by land for over 850 miles, which is the most stable system. This geographical situation allows the build up of a strong air circulation pattern which is virtually independent of atmospheric events elsewhere. The present displacement of the ICZ a few degrees to the north is believed to be due to this powerful circulation system in the Southern Hemisphere. Were the Antarctic ice cap to melt appreciably, the release of large volumes of cold fresh water would push the ICZ even farther to the north, compressing the geographical zones northwards and altering the climate of Europe and North America, possibly melting all of the ice in the Arctic Ocean. Increasing the temperature of the tropical zones, particularly those of the Southern Hemisphere, would tend to increase the exchange effect between the equatorial and the Antarctic region and allow the ICZ to move southwards, nearer the geographical equator.

The distribution and paths followed by near ground air currents are governed by the upper westerly jet streams which guide the depressions and anticyclones. Over the oceans there is little friction exerted on the air flow and it is here that moisture is picked up to feed the rain clouds which develop in response to convection. This moisture is precipitated from the atmospheric clouds over land or mountain chains where the wet air is cooled by being forced to rise, either as a result of mountain barriers or because of warm air convection taking the moisture-laden air up into the cooler regions. Over large mountain chains, however, a considerable deflection can be exerted on the low-level flow of the atmosphere and locally change the weather patterns. However, it is the overall large-scale weather patterns which are really important when considering survival during adverse climatic conditions. Through many years of painstaking research we understand a great deal about how the climatic processes of the world operate, but can we now use this knowledge of our climate to tell what the future holds in store? The answer depends entirely upon our interpretation of the weather patterns of yesteryear, for as surely as the present climatic processes hold the key to the interpretation of past, so the patterns of the past hold the key to the future.

The detailed study of the growth of trees and annual rings they produce during each season's growth (*dendrochronology*) has revealed much valuable information. During particularly cold spells only minimal growth occurs and this produces narrow rings, while during optimum warm, moist weather the trees grow more quickly and this produces wide growth rings. The interplay of temperature, available water, winds, day length and other local factors are all considered in deciphering why trees grow as they do. The results of this work have allowed us to delve far back into the past with surprising accuracy; as a check on this accuracy we have the detailed chronologies of Egyptian history and such valuable information as the old weather records collected by sea captains of many years ago. The ships' logs and weather records were brought home and preserved, for after all a knowledge of the weather was essential for the survival of sailing ships. Some ships took ocean temperatures when crossing the Atlantic and from the records preserved in the National Maritime Museum at Greenwich we know that during the eighteenth-century 'little ice age', the Gulf Stream flowed considerably south of its present position, suggesting that the colder polar mass and the air circulation patterns were also displaced to the south at that time.

In the United States, a careful study of the Ponderosa Pine, particularly of the bristlecone variety in the White Mountains of east central California, has allowed a great deal of data on annual growth rings to be amassed. A standard tree-ring chronology for the past 900 years has been established and an important by-product of this work has been a recalibration of carbon-14 dating. Processed by computer, tree-ring data can reveal much valuable information. When samples from California and Colorado were compared, for example, the differing weather conditions of 1849 were revealed. This was the year of the south California gold rush. The Californian trees had narrow rings which indicated poor growth and meagre water supply, whereas 700 miles away in Colorado growth rings of the same year were wide and indicated good growth. Obviously the climatic conditions were very localized during this period, otherwise the trees would have shown evidence of similar growth conditions.

Today man is steadily adding his quota to pollute the atmosphere, and the effects of this are no better seen than around some of the heavily industrialized cities of the world. It is not uncommon to see ascending columns of steam and smoke forming and feeding clouds, which slowly grow into towering cumulonimbus columns to eventually release the water with which they have been fed as rain. Mexico City, for example, lies beneath a perpetual pall of smoke, fumes and dust which creates a 'greenhouse effect', trapping the heat beneath the pollution layer which still permits solar radiation to pass through. Many scientists believe that man's own efforts are slowly changing the world's climate, for climatic changes are occurring today which in the past were only associated with high rates of volcanic activity, i.e. with natural atmospheric pollution. Industrial pollution is not the only culprit: the wholesale burning off of areas of forest, scrubland and corn stubble and the misuse of land to create dustbowl conditions, all pollute the atmosphere. It is a cumulative effect, for it is these contaminating dusts, vapours and gases which, unless steps are taken to prevent atmospheric pollution, will eventually reduce the solar energy we receive and cause cooling on a global scale, with the immediate result that the climatic zones of the Northern Hemisphere will be forced southwards nearer to the equator. The deserts of the Northern Hemisphere would similarly migrate southwards and encroach upon the savannas, drying out the soil and allowing further dust-bowl conditions to develop. In addition, the southerly movement of these climatic zones would depress the monsoons of India and Africa and result in the failure of valuable crops year after year with, inevitably, widespread famines and the death of millions of people.

While the increase in carbon dioxide in the atmosphere, amounting to some 10 per cent since the end of the nineteenth century, means that the world today may be warmer than it would otherwise have been, this will not necessarily prevent the slow inexorable approach of a period of colder climate. The 'greenhouse effect' of the carbon dioxide may simply be helping to delay the onset of the next ice age. On the other hand, whereas the increased influx of solar heat allowed by the build up of carbon dioxide enhances the circulation of the atmosphere of the world and leads to increased cloud formation, this, in its turn, cuts down the amount of solar radiation our planet is able to receive. Thus, the effect of an increased carbon content is, to some extent, nullified. A further factor in this story is deforestation: for it is green plants that remove carbon dioxide from the air and release life-giving oxygen in its place. For this reason alone, it is important that such great natural forests as those of Brazil and the Congo should be preserved, and the open ocean waters conserved.

In 1970, the savanna of the southern Sahara was subjected to a severe drought, with the result that large areas of agricultural land were laid to waste. Much of the soil removed by wind erosion was later detected with laser beams in a natural pollutant dust layer 12 miles up over the Pacific Ocean. Thousands of trees and hundreds of animals died in this drought, which was just one of a series suggesting that the northern deserts were moving south in the early and mid 1970s in response to a cooling of the rest of the Northern Hemisphere. This trend has now been reversed, but if such a drastic change in our climate were to continue for several decades, it could eventually lead to the evacuation of entire nations.

A stable climate and regular rainfall are of paramount importance in warm fertile regions such as India, where a precarious agricultural system must feed millions of people. Here, the population has waited each season for the last 9000 years for the monsoon to come and irrigate their crops. Most years they did not wait in vain: the monsoons always came, although sometimes they were late. Only in exceptional years, such as 1895 and 1899, did the monsoon not bring the expected rain. The population of India continues to grow: a few such bad years with late or poor monsoon rains could result in famine on an unprecedented scale. Look at the recent record: in 1972 one-third of the food was lost due to bad weather, while 1973 was a fairly normal year with sufficient for everyone. In 1974, the monsoon was late and when it did come little rain fell at first; later, simply torrential rains caused the terrible floods which inundated Bangladesh and parts of Burma, causing famine, disease and deprivation to the populace, while other parts of India went short of rain. During the Indus Civilization of Rajastan in north-west India, some 4500 years previously, the area was one of flourishing agriculture; in fact, it is regarded as one of the sites where farming civilization began. For over 1000 years the crops grew abundantly as each monsoon came and went with thankful regularity: then the monsoon failed and desert conditions crept in, although elsewhere in India rains were plentiful. Many other areas now require irrigation where once the seasonal monsoon would suffice.

Many thousands of years before its period of fertility in the fourth and fifth millennia BC, Rajastan suffered even worse desert conditions than those at present. This was during the last great ice advance in northern Europe and northern North America, around 18,000 years ago. As its period of fertility coincided with the climatic optimum of the current interglacial, the return of desert conditions today should possibly be regarded as evidence that the Earth is already well on its way towards the next glacial period. This matter is at the heart of the near future frozen death or flooding earthshock problem. What will happen first? Will the ice caps suddenly increase in size, or have we still to pass through a climatic oscillation, during which one or both of the polar ice caps will melt, before full glacial conditions return? Although it is certain that many marine transgressions and regressions, ice advances and retreats will be seen by mankind in the future, we cannot yet predict which of these disasters will occur next, or how soon it will occur. We must be prepared for the worst. We can easily visualize what a full glacial period would be like by comparison with parts of Greenland and Antarctica today, where the ice sheets are over a mile thick. It is a perpetual winter where the summer sunshine fails to melt the winter's snow, and where vast areas are held in the vice-like grip of millions of tons of ice. All this ice locks up great volumes of water, and as it is formed principally from frozen water vapour, it follows that the salinity of the oceans is greater during ice advances, because vast volumes of water are removed from them during evaporation and not returned. Again, when great volumes of water are locked up as ice, there is, quite naturally, a fall in sea level. This leads to the exposure of large areas of continental shelf sea floor as new land, and causes new beaches and marine platforms to be cut at lower levels. In high latitudes, the great thicknesses of snow that fall and accumulate on these exposed sea floors mean that they are quickly added to the ice caps.

Outside the ice bound areas, there are tundra conditions, yielding a cold desert with the ground frozen down to many feet; conditions such as those which occur during Siberian winters today. Mud dredged from the floor of the oceans indicates that 18,000 years ago there were extremely low temperatures throughout the world: in any new glacial period, conditions would be the same outside the actual area inundated by ice. World food production potential would be greatly reduced. As we have seen, recent research tells us that the prospects of the sudden onset of an ice age are far more terrible than had hitherto been supposed. There will not be time for nations to migrate slowly away from an advancing ice sheet: they will be suddenly caught up in the midst of it, virtually overnight. Again, while we know that man's activities could delay the advent of an ice age, we also know that, in his ignorance, he could equally well bring forward the dreadful hour. Undoubtedly, the earthshock of another ice age could be as devastating as a thermonuclear war, even without the attendant radiation hazards. It will put immense pressure on world food and agricultural resources, and on energy and economic mineral reserves.

Around the fourth millenium BC it is possible that for a short time no permanent ice cap existed in the Northern Hemisphere, but it rapidly became re-established, with corresponding changes in the climate of Siberia, Europe and North America, including a general lowering of temperatures and an increased number of depression tracks at lower latitudes. The attendant rise in world sea level during this optimum period of climatic amelioration was of the order of 12 feet. There have been several subsequent oscillations of lesser magnitude in more recent times, including one sea level rise of the order of 10 feet and several low stands during which times many now submerged forests grew. Along the southern shores of the English county of Kent, for example, we can find clear evidence of low sea levels in Roman times and in the eighth and fifteenth centuries; in the thirteenth century, on the other hand, the sea level was high, close to that of today. The rise since the fifteenth century in this area has been at the rate of about a foot per century. Between the Wash and the north Kent coast of England the land is slowly subsiding towards the North Sea basin as a result of the complex interplay of isostatic and geotectonic forces. The oscillations of world sea level caused by ice melting or ice cap growth at the poles is superimposed on this long-term downward movement. Thus, in London, the Thames Estuary and much of East Anglia, evidence for the recent sea level advance is particularly strong. Roman remains at Tilbury, for example, are now 15 feet below high water mark. We will return to this particular problem of subsiding coastlines in Chapter 8.

When, at some time in the future, a general amelioration of the world's weather results in the melting of most or all of the ice at present held at the poles, and the water at present locked up in the ice caps and glaciers of the world is returned to the oceans, a devastating earthshock will ensue. The rising sea level will rapidly inundate every low-lying area on Earth – all the major industrial cities and many others will be flooded (Plate 21). The greatest loss to mankind, however, will be the conversion of virtually all the best agricultural land in the world to shallow continental shelf seas. A rise in sea level of only 100–200 feet would be almost as terrible as a rise to the full 330 feet that is possible. The present highlands of the world – all that would remain of the land – would not be capable of sustaining the present world population. In view of man's poor state of social evolution, particularly with regard to greed, dogma and inability to coexist peacefully, it seems inevitable that the food shortages that would result from either a major ice advance or retreat would quickly spark off wars. These would almost certainly escalate into nuclear ones. Are the governments of this planet aware of all the facts? Do they realize the possible consequences of another ice age, or of wholesale or even partial melting of an ice cap and of flooding of most of the major cities of the world? In over 95 per cent of cases the

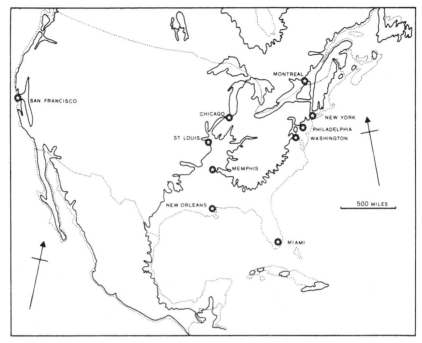

Fig. 27   Ice-melt hazard: North America. Extent of the effect of 75 per cent polar ice-melt. The chief hazard, apart from the loss of all the urban seaboards and many inland cities – several cities would actually become major ports – is the total inundation of vast areas of essential agricultural land.

answer must be no. We must be concerned about the future and what it holds in store, and in this respect it is perhaps correct to say that bad predictions are worse than none at all.

On a number of occasions during historical times man has been subjected to unusually severe winter conditions, as we have seen. These 'mini' ice-ages probably caused less inconvenience during the nineteenth century than they would in our advanced technological society. However, it is not the 'mini' ice-age with which we are here concerned. Man must now come to terms with his environment and changing weather and contemplate the consequences of a big freeze-up – a return to the glacial conditions that existed in the Northern Hemisphere some 10,000 years ago.

When the next glacial episode begins it is doubtful if mankind will be prepared for it. The adverse long-term weather forecasts which may well give an indication of what lies in store may be misinterpreted, and accepted merely as yet another severe winter; in its own way this forecast will be quite correct, if only an understatement of the case. The onset of the big freeze-up may be progressive, with snow falling steadily throughout the winter months, or the onset may be sudden and accompanied by blizzards like those which struck North America in early 1977. Whatever happens, the end result will be the same – deep, extensive snowfields throughout the Northern Hemisphere by the end of February – and only then will the difference from a normal winter become apparent, for the long-awaited thaw will fail to come and the cold weather with its occasional snow

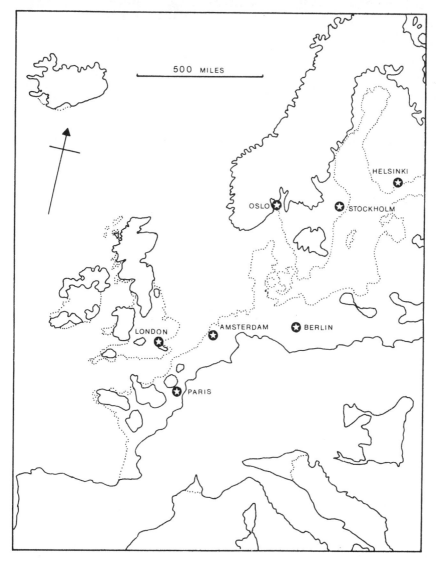

Fig. 28 Ice-melt hazard: Europe. Extent of the effect of 75 per cent polar ice-melt. Great Britain loses all agricultural land and becomes a group of small islands. The Netherlands vanish along with all urban seaboards and many inland cities.

and sleet showers will drag on and on. Exceptionally heavy snowfalls, sub-zero temperatures and galeforce winds will undoubtedly para- lyse many of our northern cities, as traffic is brought to a standstill, railway systems fail and lines of communication become broken (Plate 20).

Local authorities and organizations will be unable to cope with weather conditions of this severity, ordinary snowploughs will fail to operate efficiently and it is doubtful if sufficient of the more powerful snow-cutters will be available. Whatever attempts are made to remove the snow will be doomed to failure as the thickness of the white mantle steadily increases. Efforts to uncover roads will prove too great a task and very soon it will be apparent that it is a losing battle. In any case there will be little point in removing the snow, for its insulating property will protect the surface; snow-covered ground rarely falls below $-2°C$ and an ambient air temperature of $-20°C$ could cause severe damage to an unprotected road surface. For the same reason, houses that are covered partly by snow will require only a nominal supply of fuel to maintain an acceptable inside temperature; such energy savings could be of considerable economic value in a region of depleted fuel supplies.

As the summer of the first year passes some snow will melt during warmer periods, when temperatures may rise as high as $16°C$, while during the colder dry spells some snow will evaporate directly to water vapour without even going through the liquid stage in a process known as ablation. Agriculture will be particularly badly hit as the greatly extended winter season delays ploughing and preparation for planting, even in those areas that do become snow free. The higher summer temperatures will prove of little help to agriculture, as the level of precipitation gradually falls; the period of optimum growing temperatures will be greatly curtailed and the early onset of the next winter will prevent ripening of the crops that are able to grow. Most crops will, therefore, either fail or be late, due to the compression of the growing season.

Meanwhile, reports will be arriving from such places as the south- ern Sudan, northern Nigeria, northern China and Pakistan on the Eurasian continent, central Mexico and some of the southern states on the North American continent, telling of extensive drought and dustbowl conditions. Here, the steadily changing weather conditions will be inexorably forcing the desert zones of the subtropical high pressure regions southwards towards the geographical equator. Entire tribes will either die of starvation, or migrate south to greener pastures at the risk of mortal conflict with the natives whose land they invade.

The picture in Europe and North America by the end of the first summer will be very dismal indeed. Thousands of old people and children will have died prematurely of hypothermia, influenza and similar ailments. Fuel, such as oil, coal and natural gas, will be at a premium. Forests will become depleted and wood structures torn down as people scour the countryside in the never-ending search for fuel. Perhaps by August of the first cold year only a few inches of snow will remain in the northern temperate zone – but conditions will have been growing steadily worse and, as the second winter begins, with its near arctic conditions, both scientists and governments will come to the realization that a really big freeze-up is just beginning. By now food rationing will be in force and reserve food supplies will have been broken into as the failed crops fall far below the minimum demand. Famine, the scourge which in the past has repeatedly hit the

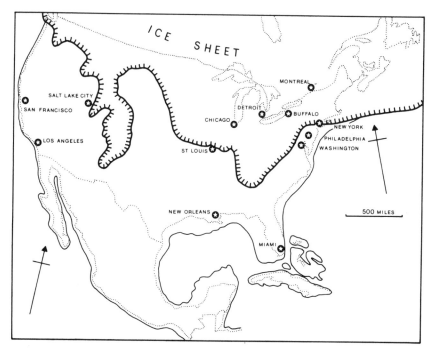

Fig. 29 Glaciation hazard: North America. The return of an ice age to the northern hemisphere could be equally as devastating as total ice-melt. All of Canada and large areas of the United States would be buried beneath many metres of ice and snow, while south of the ice cap enormous areas will be laid bare by Siberian-type tundra conditions; this will affect most areas of agricultural importance. As glaciation proceeds, the sea will withdraw and become more saline as more and more water is used to form the growing ice cap. This will leave all the major shipping ports high and dry many miles inland.

poorer and less well-developed nations of the world, will now start to gnaw at the technologically advanced western civilizations who had not considered even the remote possibility that such an earthshock could affect them. Cases of malnutrition, and especially vitamin C deficiency, will become increasingly common as fewer fresh vegetables and fruit become available.

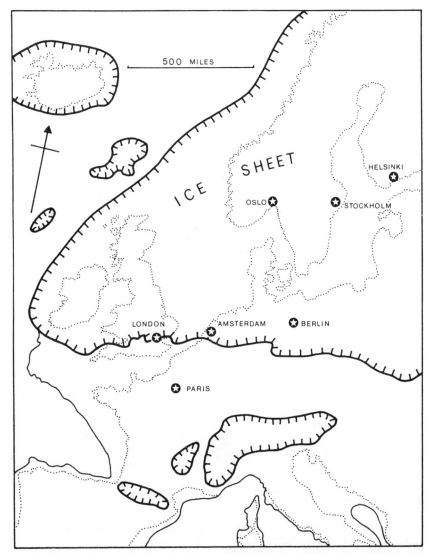

Fig. 30   Glaciation hazard: Europe. Because of its latitudinal extent Europe would suffer under the return of glaciation to a greater extent than North America. Most of the EEC capitals would fall before the onslaught of an expanding polar ice cap as winter snow falls exceed the summer thaws. The associated withdrawal of the sea to near the edge of the continental shelf would expose both the sea floor and lowland south of the ice cap to harsh tundra conditions and destroy valuable agricultural land.

The second winter will bring even heavier snowfalls to add to that remaining from the first, and so provide the first steps on the course to glaciation. Assuming that 20 per cent of the snow which falls each year remains unmelted, then at a rate of only 3 feet of snowfall per years, the permanent cover will reach a thickness of 6 feet in 10 years. Greater thickness would be expected further north and lesser values near the southerly margin of the permanent snow field.

Hydroelectric power stations will fail as their water supply becomes frozen. Those powered by fossil fuels will be dependent on their supply of coal and oil via established routes of communication and once these fail so will the power stations. Atomic power stations will probably last a little longer, at least until their supply of cooling water freezes up. Their warm effluent may delay freezing of the reservoir for a time, but eventually that too could fail. Lowering of the sea will occur as more and more water becomes locked up in snow and ice. Once this happens, seaports will become unusable, and with no alternative fuels available as the already depleted forests are decimated by the remaining survivors or buried by snow, the collapse of civilization as we know it will be imminent. Those who do not die of starvation will certainly die of exposure, and others, who have tried to migrate south in search of warmer climes, will inevitably have to face staggering cultural and social problems as they become the minority race in what must become a largely coloured world. In an earthshock such as this it is the white man of the Northern Hemisphere who will be brought to the edge of extinction, and possibly beyond.

The prospect of the loss of our northern lands through the onset of a big freeze-up is frightening enough, but we must also consider the consequences should either of the polar caps decline in size and eventually melt completely. It has been calculated that should all of the ice in the Antarctic melt then the average sea level throughout the world would rise by 330 feet, sufficient to cover the dome of St Paul's Cathedral in London. There can be no doubt whatever that such an earthshock would be more devastating than 'frozen death' for, without exception, nearly all the important agricultural land of the world would be totally destroyed by sea water. Most of the major cities of the world – New York, London, Paris, Tokyo, Sydney, Melbourne and Rome, for example – would be irretrievably drowned beneath many fathoms of ocean. Only cities such as Peking, Lima, Kampala and possibly Moscow would survive (although there is some uncertainty about Moscow). Melting of the Arctic ice cap would have almost as devastating an effect and it is therefore pertinent to consider what would happen should this vast region of less stable ice melt, with or without complete melting of the Antarctic ice.

Man's desire to slake the thirst of the deserts to help meet the ever-increasing food demands of the world's exploding population has already resulted in towing large blocks of ice from the Antarctic to parts of South America. Now it is suggested the same could be profitably done for Saudi Arabia and neighbouring countries. In moderation, such use of our ice resources would not be dangerous, but some scientists and engineers are quoted as suggesting that the ice caps should be broken up and melted with the use of nuclear or thermonuclear devices. Apart from radiation hazards, the sudden release of large fragments of ice into the Antarctic Ocean would considerably cool it, pushing the ICZ to the north. The resulting displacement of the world's climate could ultimately cause extensive melting of the Arctic ice. In addition, there is the possibility of neither of the ice sheets being able to recover quickly enough from this kind of treatment and the probability of widespread fracturing and break up of an entire ice continent due to the sudden unloading of large artificial ice floes from its margins.

Thus, the problem of melting proceeding beyond a critical recovery point and precipitation at the polar zone being unable to keep pace with thawing could arise naturally or be accelerated by the activities of man. In either event it could lead to the total melting of the Arctic ice. It has happened before and could therefore happen again. The onset of such an earthflood would be slow and insidious. The first effect would probably be noticed in places like London and the Netherlands where, even now during inclement weather, North Sea surges cause terrible flooding and loss of life and property. In the event of extensive melting of a polar ice cap over, say 200 years, London would flood during the first 6 years; even if the ice took 1000 years to melt, London would flood in 28 years, perhaps even less. Authorities and scientists would soon recognize the problem and quickly order the construction of dykes and polders to keep back the sea, but by that time it would be a hopeless situation, and just as King Canute failed to hold back the tide, so could the authorities be doomed to miserable failure. The Netherlands, the Po Valley in northern Italy, northern Germany, Denmark and the northern shores of Soviet Russia would all sink below the advancing tide of rising water. Parts of Iraq, Saudi Arabia, East Pakistan, China and large areas of South America – the Amazon Basin, Argentina and Venezuela – would be particularly badly hit. In North America Canada would lose large areas of her northern lands, while, in the United

14   St Pierre in its heyday – a thriving prosperous port, often referred to as the little Paris of the Caribbean – but a town that was destined to die.

15   St Pierre after the 1902 disaster. All the 28,000 inhabitants of the town, save two, perished in the holocaust. The scene is not unlike that produced by an atomic bomb, the difference being that the eruption was nearly 1000 times more powerful.

16   Victim of the 1902 earthshock of Martinique, Lesser Antilles. The glowing clouds kill by asphyxiation, burial by falling masonry, severe burns and cadaveric spasm. Here, the sudden heat blast has caused the abdomen to swell and burst, releasing the contents.

17   In the 1923 earthquake in Japan, many of the 143,000 people killed died from events subsequent to the main earth tremors. Over 35,000 people were burnt to death in this one place in Honjo, Tokyo.

18 Plaster cast of one inhabitant of Pompeii who did not escape the AD 79 eruption of Mount Vesuvius, Italy. After burial by volcanic ash the corpse decomposes to leave a cavity mould. Archaeologists excavating the site fill these cavities with plaster of Paris to obtain such casts.

19 Tethered animals were unable to escape the ashfall and glowing avalanches that followed the hail of volcanic ash on Pompeii. This dog, like the man in the previous photograph, probably died of asphyxiation.

20 The return of another ice age would initially make scenes like this commonplace for most of the year. Even the short hot summers would fail to melt all the snow as northern latitudes progressively became snow and ice capped.

22 Halley's Comet was clearly visible in the English sky in February AD 1066. The Bayeux Tapestry shows an astrologer telling King Harold that it is an omen of misfortune.

21 Polar icemelt would inundate all low-lying areas of the world. This photograph of the 1977 floods in Lyons, France gives some idea of the consequences of large-scale flooding.

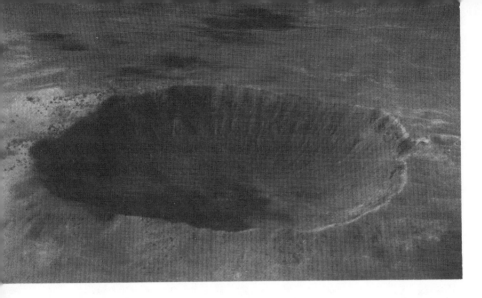

23   Barringer (or Meteor) Crater in Arizona, USA. A great circular pit ¾ mile across and 600 feet deep made 25,000 years ago by the impact of a high speed missile from space striking the Earth.

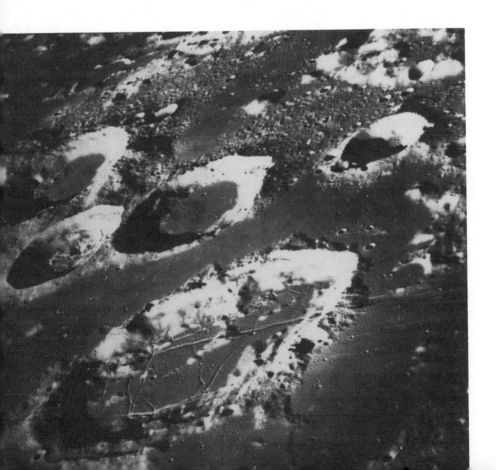

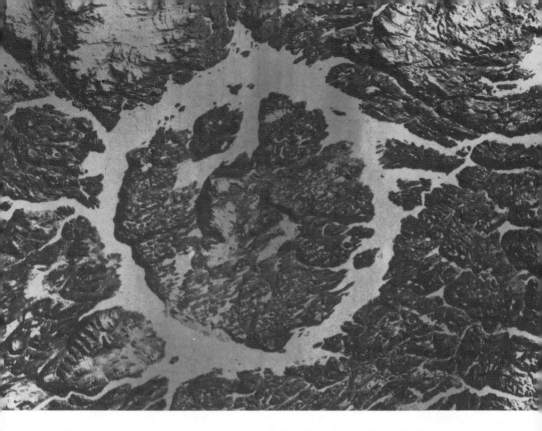

25 Lake Manicougan, in north-east Quebec, Canada is an astrobleme 37 miles across. Astroblemes or 'star-wounds' are the ancient scars on the Earth's Crust left by collisions between our planet and smaller astronomical bodies travelling at high speeds through space.

24 Four huge impact craters on the Moon. The largest is Goclenius, 47 miles across and 6100 feet deep. Crossed by a conspicuous lunar rill, it occurs near the edge of Mare Fecunditatis. Beyond Goclenius the line of smaller craters are, from left to right, Magelhaens A (20 miles across, 5700 feet deep), Magelhaens (24 miles across, 6428 feet deep) and Gutenburg D (12 miles across, 4133 feet deep). Behind them is Colombo A (27 miles across, 5970 feet deep) and the edge of the Mare Nectaris. The two small craters on the right hand edge of Colombo A are Colombo G and H, 7 and 4 miles across respectively.

26 Scientists collecting gases on the flanks of an erupting volcano. Analysis of such gases yields valuable information about the course of volcanic eruptions and could be useful, together with geophysical measurements, in predicting dangerous eruptive phases.

States, most of the Mississippi valley would be inundated with the consequent loss of such states as Louisiana, Mississippi, much of Texas including Houston, most of Alabama, and large areas of Arkansas and Tennessee. On the east coast, Florida would disappear; Georgia, Carolina and Virginia would lose their seaboard ports and cities, while Boston, Washington and New York would be inundated. Unlike the partial loss of the Northern Hemisphere by glaciation and the consequent southwards migration of the climatic zones, therefore, the loss of these vast areas of vital agricultural land would be sufficient to bring about famine on a scale as yet unimagined. Such an earthshock would so change the pattern of civilization as we know it that the evolution of *Homo sapiens* could enter an entirely new phase. In fact, many forms of animal and plant life could be brought to extinction or, at best, be set upon new evolutionary trends.

7 Extraterrestrial Bombardment

Above the clustered roofs, low in the southern
sky, they saw a great star with a three-fold train
of fire . . . Duke William said 'It is the sword
of God drawn against King Harold. Michael and his Angels
fight for us.'

Hope Muntz, *The Golden Warrior*

We now know for certain that some 25,000 years ago a very large meteorite struck the Earth, forming a huge crater nearly ¾ mile across in the desert of north central Arizona. Discovered in 1891, the great circular pit 3937 feet in diameter, 600 feet deep and with a rim rising 130–160 feet above the surface of the surrounding desert, was originally called Coon Butte, but is now known as Barringer or Meteor Crater (Plate 23). As early as 1894, the meteoritic character of the fragments of iron found around the crater had been established, but many scientists persisted in the belief that it was of volcanic origin. The possibility of the Earth being a target for large extraterrestrial missiles was unacceptable to orthodox science until well into the twentieth century. A mining engineer, D. M. Barringer, spent 25 years trying to prove that the crater had been caused by meteorite impact, but general acceptance of his hypothesis only came after other, undoubted meteorite craters were discovered at Henbury in Australia in 1931.

Fragments of iron meteorite up to 100 lb in weight and amounting to over 20 tons in total have been found scattered over the Arizona desert around the crater for distances up to 6 miles. At the bottom of the crater are fossiliferous lake deposits, 70–90 feet thick, dating from the penultimate stage of the Pleistocene glaciation in North America: Barringer Crater must, therefore, be around 25,000 years old. Ancient Permian sandstones and limestones surround the crater and are inclined outwards in all directions. Fragments and blocks of rock, mainly limestone, are scattered over the rim, in the crater itself and for some 6 miles around; some of these rocks weigh as much as 4000 tons. The dissenting hypotheses suggesting the origin of the crater as a result of volcanic explosion or cavern collapse, against which Barringer had to fight so hard, have now been completely discounted, for the circle of the crater is too perfect and the concentration of meteoritic iron around the pit cannot be explained satisfactorily by any other theory of origin. For a long time Barringer thought that a large iron meteorite lay buried beneath the crater. Several hundred thousand dollars were wasted in drilling exploratory boreholes but after passing through crushed and metamorphosed Permian sandstones, undisturbed strata were encountered beneath the crater at around 620 feet. No large body of metal was found and at one time this cast more doubt on the meteoritic origin of the crater until it was realized that the evidence indicated that on contact with the Earth's surface the meteorite had not buried itself but exploded. Barringer Crater is, in fact, a large meteorite impact fragmentation crater, young enough still to retain its associated iron meteorite fragments. It is thought that the meteorite that caused the depression must have weighed many thousands of tons, possibly as much as a million tons, but was entirely fragmented and extensively vaporized on impact. The enormous impact explosion caused disruption and partial melting of the layers of sandstone and limestone beneath the desert surface to a depth of several hundred feet. Small glassy spherules of fused rock are found on the surface around the crater for several miles and these are thought to have been splashed outwards from the crater. Other signs of intense impact are the formation of rock flour (finely pulverized rock dust) and the presence of certain distinctive silica minerals that are thought to form only under conditions of extremely high pressure. It has been estimated that the impacting mass must have exploded with a force equal to at least 5 megatons of TNT.

Every day of our lives we are in danger of instant death from small high-speed missiles from space – the lumps of rocky or metallic extraterrestrial debris which continuously bombard the Earth. The chances of anyone actually being hit, however, are very low, although there are recorded instances of 'stones from the sky' hurting people, and numerous accounts of damage to buildings and other objects. At night this extraterrestial material can be seen as 'fireballs' or 'shooting stars', burning their way through our atmosphere. If they survive their fiery entry into the Earth's atmosphere so that some part of their solid substance reaches the ground they are termed *meteorites*;

most, on reaching our atmosphere, become completely vaporized and scientists call these *meteors*.

The height above ground at which these objects become sufficiently heated to be visible is estimated to be about 60–100 miles. An iron meteor as small as the head of a pin can shine with a light equal to that of a bright star. Surfaces of fresh snow have sometimes been seen to be covered with minute drops of melted iron after a group of shooting stars has crossed the sky. Meteorites that have fallen on buildings have sometimes ended their long lonely space voyage incongruously under beds, inside flower pots or even, in the case of one that landed on a hotel in North Wales, within a chamber pot. Before the era of space exploration it was confidently predicted that neither men nor space vehicles would survive for long outside the protective blanket of the Earth's atmosphere. It was thought that once in space they would be seriously damaged as a result of the incessant downpour of meteorites falling towards our planet at the rate of many millions every day. Even the first satellites showed that the danger from meteorites had been greatly overestimated by the pessimists, but although it has not happened yet, it is certain that one day a space craft will be badly damaged by a meteorite.

The greatest single potential danger to life on Earth undoubtedly comes from outside our planet. Collision with another astronomical body of any size or with a 'black hole' could completely destroy the Earth almost instantly. Near misses of bodies larger than or comparable in size to our own planet could be equally disastrous to mankind as they might still result in total or partial disruption. If the velocity of impact were high, collision with even quite small extraterrestrial bodies might cause catastrophic damage to the Earth's atmosphere, oceans and outer crust and thus produce results inimical to life as we know it. The probability of collision with a large astronomical body from outside our Solar System is extremely low, possibly less than once in the lifetime of an average star. We know, however, that our galaxy contains great interstellar dust clouds and some astronomers have suggested that there might also be immense streams of meteorite matter in space that the Solar System may occasionally encounter. Even if we disregard this possibility, our own Solar System itself contains a great number of small astronomical bodies, such as the minor planets or asteroids and the comets, some with eccentric orbits that occasionally bring them close to the Earth's path. While there are considerable difficulties in assessing the likelihood of a collision between the Earth and one of the larger comets or asteroids with any degree of accuracy, the statistical estimates that have been made certainly suggest that collisions of this kind do happen and, indeed, were extremely frequent during the early history of the Earth. Another way of approaching the problem is to examine both the historical record and the geological 'record of the rocks' for evidence of past impact events. From this evidence, if it is at all representative, some assessment of the probable extent and frequency of future extraterrestrially induced disasters can be made. Extrapolating from the probable frequency with which the Earth is bombarded with very small astronomical bodies – the meteors and meteorites – it is seen that collisions with bodies of the size of minor planets might be as frequent as one every 1–50 million years (Table 7.1). This estimate is not incompatible with the fact that at least 50 large impact features – the so called *astroblemes* or 'star-wounds' – have already been identified on our continents.

Table 7.1   Approximate present frequency of extraterrestrial bombardment suffered by the Earth

Weight of objects before entry into the Earth's atmosphere	*Approximate average frequency*
space 'dust'	over a million tons filters down to the surface of the Earth each year
meteors only visible with telescopes (burnt up in the atmosphere)	over 1,000,000,000 every day
meteors visible to the naked eye (burnt up in the atmosphere)	over 500,000 every day
small meteorite *c.* 10 lb (only a few ounces reach the ground)	3 or 4 every day, but very irregular, often arriving in swarms composed of considerable numbers
5 tons (not more than about 1000 lb reaches the ground)	1 every month
50 tons	1 every 30 years
250 tons	1 every 150 years
50,000 tons	1 every 100,000 years
small planets (asteroids) with diameters measurable in miles	1 every 1–50 million years

The movements of the Sun and Moon and the slowly revolving pattern of the stars in the night sky have fascinated man since the earliest times. Many men have spent long hours of their lives staring up at the sky, and the influence of their nocturnal observations on religious belief and observances has been tremendous. The great stone circles, such as Stonehenge in England, illustrate the importance of astronomical observations in the lives of our ancestors. Unusual features of the night sky have almost always been interpreted in terms of portents and evil omens. Halley's Comet made an appearance in AD 1066 on the eve of the Norman Conquest and this bad omen for King Harold of England is recorded for posterity in the Bayeux Tapestry (Plate 22). Throughout history men have been fascinated by 'shooting stars', 'fireballs', 'thunderbolts' and 'sky stones'; they were once thought to be 'gifts from the gods' and even the smallest meteorites were collected and carefully preserved in temples. The temple of Diana at Ephesus, for example, contained a meteoritic stone which was supposed to have been thrown by the goddess. The famous Black Stone enclosed in the Ka'ba in Mecca is still an object of pilgrimage for all Islam. A large meteorite which fell in Phrygia during the third century BC was worshipped as an image of the goddess Cybele. In 218 BC an oracle declared that the Carthaginians would be defeated if the stone were brought to Rome. The meteorite was promptly taken from the Phrygian King and delivered to Rome with great pomp, where, because of its supposed assistance in the Roman victory, its worship continued for over 500 years. Early witnesses of these falling objects did not hesitate to embellish their accounts at the expense of fact and attributed meteorites to all kinds of supernatural causes. In AD 1400 an iron meteorite weighing about 235 lb fell near Elbogen in Bohemia: it was believed to represent the metamorphosed remains of a local tyrant! However, it was not until after a fall at Benares in India in AD 1798 that scientists began to ascribe the phenomenon to a natural cause.

In April 1803, the first scientific study of a hail of stones was made at L'Aigle in France. The Academy of Sciences of Paris sent an investigating committee to L'Aigle under the leadership of the physicist Biot. He reported that on Tuesday 25 April 1803, towards one o'clock on a fine afternoon, an extremely bright flaming sphere had been seen moving rapidly through the air by many of the local inhabitants. Several moments later a violent explosion lasting 5 or 6 minutes was heard over an area extending at least 50 miles from L'Aigle in all directions. The explosion consisted of three or four cannon-like reports followed by a discharge resembling first a fusillade and then a deafening roll of drums. This noise came from a small elongated cloud lying about 1½ miles NNW of the town of L'Aigle. The centre of this cloud appeared stationary throughout the duration of the phenomenon, only the vapours composing it spreading out on all sides from the effect of the successive explosions. The cloud was quite high, for it was simultaneously observed as overhead by the inhabitants of villages situated more than three miles from one another. Throughout the area over which the cloud hovered were heard whistlings like those of stones hurled from a catapult and at the same time a shower of solid meteorites fell to the ground. The largest of those retrieved weighed 19 lb, the smallest only a few ounces. The total number of recognizable meteorites which fell at L'Aigle is estimated to have been around three thousand. Biot showed conclusively that meteorites were stones that fell on Earth from space and could not be dismissed as ordinary terrestrial stones vitrified by lightning, which was the common view among his contemporaries who, unlike the ancients, were loath to believe in an extraterrestrial origin.

Just over one hundred years later a clearly extraterrestrial event of much greater violence was reported from Asian Russia. The morning of 30 June 1908 was cold and clear on the Tunguska River near Lake Baikal in Siberia. Some thousands of people in the area between the Yenisei, the Lena and the Trans-Siberian railway later bore witness to the flight of a gigantic ball of fire, trailing a thick dust cloud as it moved north-westwards across the sky; it was followed by loud thunderclaps and a stupendous crash, and hit the Earth with enormous force close to the Tunguska at 7.17 a.m. local time. Earthquakes and shock waves were recorded by microbarographs and seismographs as far away as England. The explosions were heard up to 600 miles away. At distances of 100–150 miles or so, men and horses were blown over and windows shattered. Closer, people were knocked unconscious. Enormous waves formed on the rivers and many houses and walls collapsed. From as far away as 250 miles from the point of impact a huge pillar of fire was seen to spurt up from the ground to a height of more than 12 miles. An engine driver 400 miles away brought his train to a shrill halt as the track of the Trans-Siberian railway heaved and quaked in front of him. At the point of impact the forest was carbonized instantaneously. Trees were uprooted and felled radially in a zone extending some 20 to 30 miles, their fallen trunks pointing away from the impact focus. During the following nights the skies of Europe and north-west Asia were illuminated by bright dust clouds similar to those seen after the explosion of Krakatoa in 1883.

The remote location of the impact site and the intervening revolution and war years in Russia resulted in postponement of formal scientific investigation of the Tunguska events until 1927. No remains of the supposed meteorite were ever found, only fused pellets – apparently of terrestrial origin – embedded in the ground like grape shot. It was assumed, therefore, by some investigators that the meteorite that caused the damage must have been completely vaporized and exploded on impact. In the central area many small craters between 30 and 150 feet in diameter were found but there was no single extensive crater, such as is found in other accepted examples of large meteorite impact. Many hypotheses have been put forward as to the exact nature of the extraterrestrial body that could cause such devastation as occurred on the Tunguska. One hypothesis assumes that it was an extremely large meteorite of the order of several thousand tons in weight that exploded far above the ground surface. Certainly the explosive violence of the event was unlike the phenomena normally associated with small meteorite impacts. It has also been suggested that the collision might have been with a body of anti-matter or with an extraterrestrial nuclear space vehicle which crashed on Earth. While the complete lack of any radioactive fall-out effects in the immediate neighbourhood would seem to rule out the latter suggestion, a number of authorities have considered the possibilities of an anti-matter collision. From a study of the eyewitness accounts, however, the more mundane explanation – that it was merely a large meteorite which came towards the surface of the Earth at a very low angle of approach and thus was almost completely vaporized before impact – could be correct. A large ballistic shockwave would certainly have been associated with such an angle of approach and this could have been the principal cause of the widespread devastation. Another very attractive hypothesis is that the impact was that of a small comet. This hypothesis would account for the high velocity of arrival and lack of any recognizable meteoritic debris, since comets consist very largely of ice, gas and dust.

In the final report of an investigation carried out for the Committee on Meteorites of the Soviet Academy of Sciences, its Chairman, Vassily Fesenkov, announced in 1960 that a small comet several miles in diameter, weighing about a million tons, was responsible for the Tunguska event. Such a comet would be only about a millionth of the weight of an average comet, but if, as appears likely, it met the Earth head on with a closing velocity of more than 25 miles/second, the energy associated with such an impact would easily be sufficient to cause the devastation seen at Tunguska. In any event, whatever actually occurred in Siberia in 1908, it is certainly the largest observed and scientifically recorded impact in recent history. With a change of only 4 hours 47 minutes in the timing of the collision, the impact area could have been the city of St Petersburg (now Leningrad) and the effect on this city would have been as devastating as the explosion of several hydrogen bombs: recent estimates of the energy released in the Tunguska event compare it to a 30 megaton explosion. Thus, one of the questions we must ask ourselves is whether, if an unexpected extraterrestrial collision of similar violence were to occur today, destroying say a city such as Washington, Paris or Moscow, might not a massive retaliatory nuclear holocaust be precipitated inadvertently before the real cause of the devastation was appreciated?

Today, there can be no doubts regarding the extraterrestrial origins of the Tunguska missile and the iron meteorite that excavated Barringer Crater in Arizona. Many similar craters, all accepted as meteorite impacts, can be seen elsewhere in the world, one of the most spectacular being Wolf Creek Crater in the Australian desert, which was discovered from the air in 1947. Another crater-like feature, of about the same age as Meteor Crater in Arizona, is occupied by Chubb Lake, at Ungava in the far north of Canada, a pale blue lake in a perfectly circular depression 11,500 feet in diameter. Some 50 or more craters have now been identified throughout the world that can be ascribed to quite recent meteoritic origin. From the

Table 7.2 Some certain (authenticated) meteorite impact craters on earth

Name and location	Size (largest diameter)	Number of craters
Barringer or Meteor Crater, Arizona	3937 ft	1
Lonar, India	3600 ft	1
Wolf Creek Crater, W. Australia	2789 ft	1
Aouelloul, Mauritania	820 ft	1
Boxhole, NT, Australia	574 ft	1
Odessa, Texas	551 ft	3
Henbury, NT, Australia	492 ft	14
Oesel Craters, Estonian SSR	361 ft	7
Wabar, Saudi Arabia	295 ft	2
Campo del Cielo, Argentina	230 ft	9
Sikhote Alin, Primorye Ter, Siberia, USSR	87 ft	22
Dalaranga, W. Australia	69 ft	1
Haviland, Kansas	36 ft	1
Plus numerous lesser and more ephemeral craters		

evidence of these craters alone, there can be no doubt that we live on a bombarded Earth, travelling like the crew of an unarmed space ship through a continuous and murderous broadside from afar.

Elsewhere in the Solar System, collisions with and between stray bodies such as meteorites and asteroids are thought to have played an important part in shaping the surface of the planets and their satellites. Recent photographic exploration has shown that Mercury, Mars

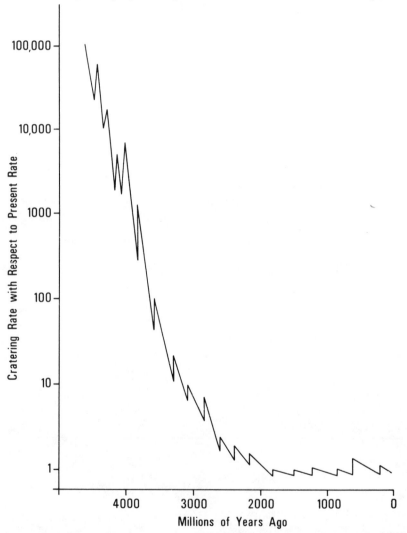

Fig. 31 The decline in the rate of cratering of the Earth, Moon and nearby planets. Intense bombardment occurred during the growth stage around 4600 million years ago. From that time on, the cratering rate declined rapidly but with occasional reversals of this trend as larger streams of uncaptured planetisimals were encountered and swept up. The peak around 4000 million years ago corresponds to the maria events on the Moon. The cratering rate fell to around its present level soon after 2000 million years and since then has remained roughly constant.

and the Moon are all densely cratered in part. On the Moon most of the more obvious cratering events appear to have occurred between 3000 and 4400 million years ago, and it seems likely that collision on a large scale was much more common in the early history of the planets. In addition, it has been argued that some of the structural features preserved in the very old Precambrian shield areas of the Earth's Crust are the relics of very large ancient impact scars, formed during this early period of intense collision. Nevertheless, space exploration of the Moon and our neighbouring planets confirms the continuation of impacts and cratering to the present day, albeit at a lesser rate.

What causes this perpetual bombardment and where do the missiles come from? If we look at the Solar System as a whole, we find that besides the nine major planets there are enormous numbers of small bodies, ranging in size from barely visible minute dust particles to the asteroids or minor planets, which may be some hundreds of miles in diameter. Altogether these minor members of our Solar System make up only a very small part of its total mass, but, as a group, they are of disproportionate interest as the main cause of phenomena such as comets, shooting stars and meteorite falls on Earth and as the source of the much larger planetisimal objects which bombard the Earth at rarer intervals. While the nine major planets of the Solar System all orbit the Sun in the same direction and in almost the same plane (the ecliptic), the minor bodies that we are interested in show more irregular motions; they may have eccentric or highly-inclined orbits, many of them not spherical in shape. The parabolic or hyperbolic trajectories followed by some of the comets suggest, in fact, that objects from outside the Solar System may be quite frequently encountered by the Sun and its family on their voyage through space. Nevertheless, most of the space debris that hits the Earth is most probably from within the Solar System.

The smallest independent particles within the Solar System form the so-called 'zodiacal light', a dim light which may be seen in a moonless night sky, emanating from the reflection of the Sun returned by millions of small dust particles in space. *Comets* are distinct astronomical bodies of low mass that roam the icy edges of the Solar System, way beyond the orbit of the furthest planet Pluto. Astronomers estimate that there are some 100,000 million or more comets, orbiting not only in the flat disc of the planets, but also in a spherical halo surrounding the Solar System and reaching out 10^{12} miles or so towards our Sun's neighbouring stars. Relatively few comets have ever visited the central regions of the Solar System around the Sun where astronomers can study them, but those that have reveal the astonishing fact that most of them are merely accumulations of frozen gases and dirt, no more than a few miles in diameter, with a density not much less than that of water. Away in space a comet has no tail: this only appears when a comet approaches the Sun, because solar energy radiating from the Sun vaporizes and expands the outer layers of the comet to form a swollen head (called the coma) and then drives some of this material away to form a streamer of incandescence pointing out towards space. While not weighing even a millionth of a millionth as much, the whole volume of a comet may then be inflated to occupy a space larger than the Sun itself. If they come still closer to the Sun, streams of solar particles can disrupt the comet, creating explosive pockets of gas in its spongy substance and stretching its tail out even more. The great comet of 1843 had a tail that streamed out more than 500 million miles. Halley's comet, which caused such consternation in AD 1066, returns to visibility every 76 years and is so brilliant that it has been recorded in the annals of Chinese and Japanese astronomers on each of its appearances but one since 204 BC. It is one of the largest comets known: its next reappearance will be in 1986. Edmund Halley, the English astronomer (1656–1742), was the first to state that most visible comets are members of the Solar System travelling in elliptical orbits. The majority of them are probably original members of the Sun's family which have orbited in and out of the central region of the Solar System from time immemorial. It is most likely the gravitational influence of a passing star which causes one of them to leave its cold distant realm and penetrate to the heart of the Solar System. Once in amongst the planets, a comet may simply pass unscathed and rocket back into its own realm or, by chance, the gravitational influence of a major planet may pull it into an orbit that brings it repeatedly close to the Sun.

The fate of comets that play too often with the Sun's fire is illustrated by the short and wild career of Biela's comet, which was first noticed hurtling in from space in the year AD 1772. It then reappeared in the Sun's vicinity at regular six-and-a-half year intervals. On its 1846 appearance it abruptly became two comets moving side by side. It put in one more appearance in this double form in 1852 and then vanished. Astronomers were still looking for it 20 years later when the whole of Europe was suddenly treated to a great pyrotechnic shower of meteors burning up as they entered the Earth's atmosphere. The rain of cosmic sparks increased as it moved westward and by the time it reached England people could see a hundred blazing meteors a minute. Over the Atlantic the display gradually diminished, failing somewhere over the Americas. Careful calculations have since proved that the meteors were really the remnants of Biela's comet crossing the Earth's orbit just in time to meet the Earth. It was lucky that Biela's comet did not strike the Earth when it first appeared in the inner part of the Solar System, because before it has been inflated, eroded and broken up by the influence of the Sun, a full-sized comet can pack a really substantial punch. A cometary nucleus hitting the Earth would release the kinetic energy of an object weighing millions of tons travelling with a velocity of anything up to several hundred miles per second – and this energy release would take the form of a colossal explosion. Once inflated by closeness to the Sun, however, both the head and the tail of a comet are transparent against the background of the stars, made up as they are of a distended lattice of dust particles, ice and gas. The density of the material in the tail may be so low that 2000 cubic miles of it are equivalent to one cubic inch of air at the surface of the Earth! Even if the Earth were to pass directly through such a comet's tail – as happened with Halley's comet in 1910 – only a tiny part of the cometary gases would impinge on the Earth, showing no noticeable effect on the atmosphere.

Asteroids are the largest of the minor bodies of the Solar System; they are irregularly shaped, varying in size from several hundred miles in diameter to a few hundred feet. Some 2000 of the larger asteroids have been identified, but there are many more present in the Solar System. They are usually thought to represent a large planet which once existed beyond Mars and which at some time disrupted to form thousands of small, irregular bodies; some, however, may simply be fragments of the primary material of the Solar System from which the other larger planets originally formed by a process of accretion. A major belt of asteroids lies between Mars and Jupiter, but the mass represented by this belt is probably altogether less than 5 per cent of the mass of the Moon.

The first asteroid of any size was recognized by the Italian astronomer Giuseppe Piazzi on the very first night of the nineteenth century – he watched it for 41 evenings and then lost it in the twilight as it moved into the Sun's area of the sky. All over Europe other astronomers searched in vain for it. It is said that during the course of a battle Napoleon discussed with one of his officers what name to give it if it should be rediscovered. The German mathematical genius Karl Friedrich Gauss, indulging his weakness for astronomical arithmetic, gave up all his other work and from Piazzi's scanty observations reconstructed the lost body's orbit. With Gauss' information other astronomers were able to examine the sky in the right place and, sure enough, they discovered the mislaid minor planet. Piazzi christened it Ceres and later measurements have revealed it to be an orbiting mass of jagged rock 480 miles in diameter. Ceres was both the first to be discovered and the largest of all the asteroids. Others were discovered in quick succession – Pallas, 300 miles wide, in 1802, Juno, 120 miles across, in 1804 and Vesta, 240 miles across, in 1807. A new 400-mile wide asteroid called Chiron, orbiting between Jupiter and Saturn, was discovered in 1977. There must be many more of these minor planets still to be discovered. Today we think that about 30,000 sizeable asteroids exist, ranging from substantial three-dimensional bodies like Ceres down to small flying masses of rock like Icarus, which is only a mile in diameter.

The number of smaller asteroids, the size of boulders, cobbles, pebbles and even sand grains, is estimated in the billions. Some 1600 have been watched carefully enough for their orbits to be plotted and predictions made about their future whereabouts. Of these, every one revolves around the Sun in the same west to east direction as the Earth and the other planets. Since they mainly move in a broad band situated between Mars and Jupiter, it is the enormous planet Jupiter which controls their motions. Occasionally Jupiter's predominant pull sends one of the asteroids on a series of orbital voyages down towards the Sun or up towards the outer planets. Icarus' present orbit takes it very close to the Sun and the inner planets; on its last passing of the Earth in 1968 it was only 16 times as far away from us as the Moon. Hidalgo whirls in and out as far as Saturn; Eros, a cigar-shaped mass of rock 15 miles long and 5 miles wide, which tumbles end over end as it circles the Sun, can come within a scant 14 million miles of the Earth on its present orbit. Amour, Apollo and Adonis can come even closer. In March 1978 an asteroid one mile across belonging to the Apollo group passed within 8 million miles of the Earth. In 1937 Hermes (also about a mile in diameter) came so close that, to astronomers who tried to follow it, it was like the fly-by of a jet aeroplane. At its closest it was only 400,000 miles distant, just twice as far away as the Moon. Eight asteroids in this size range move in orbits which cross the Earth's path, so it is clear that asteroids or small planets occasionally do come close enough to the Earth to make collision a possibility.

Really large masses of extraterrestrial origin actually strike the Earth perhaps once every 1 to 50 million years. When they do, the Earth can act like soft mud and swallow them explosively into its surface or their substance can be entirely vaporized and dispersed in a gigantic explosion. Geologists have only recently begun to recognize the wounds which the impact of large meteorites and asteroids inflicts upon the Earth. It seems likely from the evidence accumulating so far that only the shield of the atmosphere and oceans, and the healing power of erosion, sedimentation, mountain-building and vegetation, have kept the Earth from being as pock-marked as the Moon.

We know from numerous direct scientific observations that the Earth is constantly bombarded by large numbers of extraterrestrial bodies of small size – the meteorites – which enter the Earth's atmosphere in thousands every day. Most are burnt out as meteors long before they reach the surface and only about 10 meteorites are actually recovered every year from the surface of the Earth. Meteorites are particularly fascinating since, apart from the lunar rocks brought back to Earth by the Apollo and Luna missions, they are the only rock samples from interplanetary space that can be studied directly by Earth scientists. Most meteorites are thought to be related in origin to the asteroids, being smaller pieces of the same parental body or bodies, fragmented by collision or near collision and thrown into orbit as showers of smaller objects. They are of great interest to the geologist as they may include representatives of the sort of material that exists deep below the crust of the Earth, in its Core and Mantle, and fall roughly into two categories: the iron and the stony meteorites.

Table 7.3 Some fireballs and meteorites reported in 1976

Date	Locality	Description
18 Jan.	Bristol, Virginia, USA	A bright fireball visible for 15 seconds at 2030 EST was accompanied by a loud detonation that shook windows. A smoke trail remained for several minutes.
28 Jan.	Gujurat, India	A bright flash and several loud detonations at 2100 GMT. Several fragments of a stony meteorite, weighing a total of 10 lb, were recovered.
11 Feb.	USSR and Finland	A bright fireball travelling SE observed around 1550 GMT over a large area, including Tallin, Leningrad, Kalinin, Novgorod, Pskov.
8 Mar.	Kirin Province, Peoples Republic of China	A bright fireball travelling SW at 8.5 miles/second exploded with a tremendous

Date	Locality	Description
		detonation over Kirin Province at 1502 (local time). The ensuing meteorite shower was one of the most extensive falls of stony meteorites in history. Fragments were scattered over an area of 200 square miles. More than 100 were later collected. The heaviest weighed 1¾ tons and is the largest stony meteorite ever recorded. It fell in Huapi Commune in Yungchi County, breaking through 6 feet of frozen soil to penetrate 21 feet into the ground. The impact scattered clumps of earth over 300 feet from the crater, which was more than 10 feet deep.
21 Apr.	Cornwall and Devon, England	A bright fireball travelling NE seen by observers up to 200 miles apart.
21 Apr.	USSR	A brilliant fireball moving N to S at a high level lit up the ocean surface in the vicinity of Lyra at 0821 GMT.
5 June	Czechoslovakia	At 1543 UTC a bright fireball visible in the daylight sky was observed to break into two main pieces.
6 June	England	A spectacularly brilliant fireball was seen over the east coast of England from Margate to Lowestoft at 2134 UTC. For 4 seconds as it travelled E over Felixstowe in Suffolk, the fireball was trailing, with many fragments breaking off. For the last seconds of visibility no trail, fragmentation or sparks were seen, only the main body of the fireball. The colour was basically white, but reds and blues could be seen where fragmentation was taking place. 2½ minutes after the fireball was observed, several thunder-like detonations were heard.
18 July	Deephaven, Minnesota, USA	At 0225 UTC a bright green fireball was seen for about 2 seconds falling straight down to the NNE from Deephaven. The long coloured tail glowed with equal intensity for the full length of the flight. (45° to 10° above the horizon.) The ball appeared about half as big as the full moon, but there was no sound.

Date	Locality	Description
24 Aug.	California, Nevada and Arizona, USA	A brilliant greenish-white fireball travelling from S to N was visible for 5–10 seconds around 2325 PDT over portions of western USA. When first seen it was some 90 miles high: after falling to around 8 miles above the ground it exploded into a multicoloured array of fragments most of which probably fell in the heavily forested area of Yosemite National Park. People in the park and nearby areas reported a flash bright enough to awaken sleepers, a sonic boom and a smoke trail seen by some observers.
3 Sept.	Southern California, USA	A coloured fireball that appeared as big as a quarter to half the size of the full moon, crossed the sky of southern California from E to W around 0815 PDT.
20 Sept.	New Mexico, USA	A white fireball was seen for 2–3 seconds around 2007 MDT from the Sacramento Peak Observatory. A bright blue flash, brighter than lighting, occurred just after its disappearance behind clouds. A smoke trail was visible behind the fireball.
27 Sept.	Central America	A bright fireball was seen for approximately 15 seconds over El Salvador, Honduras, Nicaragua, Guatemala and nearby Mexico at around 2050 local time, followed two minutes later by a sonic boom.
4 Oct.	Beaufort, North Carolina, USA	A bright green fireball, as bright as the full moon, was seen for about 2 seconds travelling WNW to ESE around 1901 EDT.
10 Oct.	Poland	A brilliant fireball travelled E to W across south-western Poland at 1730 local time.
13 Oct.	Rwanda	At 1630 local time two strong explosions followed by a noise likened to the passage of jet aircraft were heard for up to 15 miles from Ruhobobo in Ruhengeri Prefecture. Shortly afterwards, a small meteorite weighing over 1 lb struck the ground only 6 feet from two people standing in the village.

Date	Locality	Description
19 Dec.	Western USA	A reddish-orange fireball with a blue tail was seen in Nebraska, Wyoming, Colorado and New Mexico around 1810 local time moving NNW – SSE.

The *iron meteorites* consist essentially of an alloy of nickel and iron, unlike anything found in terrestrial or lunar rocks; they appear to have been formed in a largely reducing environment, where nickel and iron occurred mainly in the metallic state, and are thought to be very similar in composition to the material forming the Core of the Earth. The textures found in the minerals making up the iron meteorites can only have been formed during a period of very slow cooling from a molten state, such as would occur in the interior of a larger asteroid or planet. Iron meteorites tend not to fragment as they travel through the Earth's atmosphere, and are usually found as much larger bodies than the stony meteorites. A meteorite weighing 14.5 tons was found in Oregon, for example, two weighing 11 and 17 tons

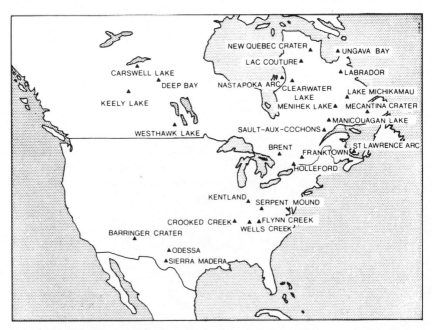

Fig. 32 A map showing the location of some of the undoubted meteorite impact craters and possible astroblemes known in North America. Positive identification of astroblemes is often difficult but as exploration of the Earth's Crust proceeds the number of structures being referred to this cause increases annually.

respectively were found in Mexico, one weighing 34 tons in Greenland, while a problematical block estimated as around a million tons has been reported at Adrar in Mauretania. The largest undoubted iron meteorite is the Hoba meteorite from South Africa, which weighs over 60 tons. The two largest stony meteorites known are a mass weighing one ton which fell in Norton County, Kansas, USA in 1948 and another weighing 1¾ tons which fell in Yungchi County in the People's Republic of China in March 1976.

The *stony meteorites* often contain small spheroidal aggregates of minerals, unlike anything found in terrestrial rocks; stones containing these spherules are known as *chondrites*, while those without are known as *achondrites*. Chondrites are by far the most commonly occurring meteorites, making up over 80 per cent of all meteorite finds. They are composed largely of olivine, pyroxene and nickel-iron, and can be compared in some ways with plutonic (i.e. deep-seated) basic magmatic rocks such as peridotite, found in the Earth's Crust. A few chondrites contain considerable amounts of organic compounds of extraterrestrial origin; these are the so-called carbonaceous chondrites and they consist mainly of serpentine together with nickel silicates and sulphites. The achondrites form a much smaller group of meteorites, usually much more coarsely crystalline than the chondrites, and showing similarities in texture and composition to terrestrial igneous rocks such as dolerite or basalt. Large stony meteorites are rarely found; few weigh more than 1000 lb, and the majority far less than this. They seem to have formed smaller masses in space and are more often fragmented during their journey through the atmosphere and on reaching the surface of the Earth. A small number of meteorites – the stony-iron meteorites – contain approximately equal silicate and nickel-iron fractions. The wide variety of chemical and mineralogical types found in meteorites suggests that they could not all have originated from a single parent body, but rather from a series of asteroidal bodies of varying size. Most chondrites show evidence of variable reheating to temperatures of 800°C or higher – the exact degree of reheating was probably controlled by the size of the parent bodies. Other chondrites which show no evidence of reheating were probably derived from smaller parent bodies. Age determinations on meteorites show a common age of about 4600 million years, and it is thought that all the major events in the genesis of the meteorite parent bodies from the original ancestral solar cloud took place in a relatively short interval of time, perhaps a few hundred years. Over the last ten years or so, certain scientists think they may have found infinitesimal fossils in meteorites, or, at least, what some cautiously describe as 'organized material'. If they are right, this implies that life existed elsewhere in the Universe long before the Sun and its planets were born and may have been transplanted to Earth from outside the Solar System.

As mentioned before, a meteorite fall may be accompanied by a brilliant fireball in the sky and sound effects which have been described as cannon fire, thunder, the tearing of calico or the passage of an express train over a viaduct. From time to time the Earth goes through periods of more intense meteor activity; these provide spectacular 'shooting star' displays in the night sky, apparently radiating from specific points, but very rarely do these meteor showers result in any material falling to Earth, as they are produced by clusters of small bodies which burn up on entering the atmosphere. Only a few of the fireball meteors that are observed every year actually drop meteorites. By night these meteors appear as brilliant balls of fire, brighter than the moon and with a path extending for perhaps hundreds of miles. Meteorites normally hit the Earth at a relatively low velocity; this is due partly to the braking effect of the atmosphere, and partly to their normally slow speed – in theory this cannot exceed 25 miles per second, since any higher velocity or orbit would carry them out of the Solar System. Most meteorites are braked to such a degree by the Earth's atmosphere that they fall towards the surface at free-fall velocity; the high temperatures reached during their journey through the atmosphere are often clearly visible as a highly fused and melted exterior. The slow velocity of impact explains why even quite large iron meteorites, weighing several tons, are commonly found lying on the surface of the ground – even on soft ground they rarely penetrate to a depth of more than 3 or 4 feet. It has been calculated that meteorites weighing less than 10 tons will normally lose all their cosmic velocity during passage through the Earth's atmosphere. If they arrive at a higher than normal speed, medium-sized to small meteorites often explode and fragment either in the sky or on impact. Thus, they may or may not result in the formation of impact fragmentation craters. These are usually associated with the iron meteorites and such craters may be surrounded by small twisted masses of metal, the remnants of the meteorite. Very fast impacts are unusual, however, since the majority of meteorites, like the asteroids, orbit the sun in the same direction as the planets, with their orbits generally close to the plane of the ecliptic. Such motions tend to produce overtaking rather than head-on collisions with the Earth. The very large meteorite craters and astroblemes that scar the Earth's Crust must, therefore, represent collision with extraterrestrial masses considerably larger than 10 tons in mass which would have arrived at the Earth's surface with much of their cosmic velocity retained.

Tektites are mysterious, small glassy objects, found lying in large numbers on the surface of the Earth in a few limited geographical areas, all at fairly low latitudes. The main areas are Czechoslovakia, Texas, the Ivory Coast, Australia and South-East Asia. All those known have fallen to the Earth in geologically recent times and four separate arrival events have been identified. In composition they are unlike any known terrestrial volcanic glasses, or any known lunar or meteoritic material; they have an unusual chemical composition of high silica, comparatively high alumina, potash and lime, and low magnesia and soda. Their origin is a mystery and has been a source of generally fruitless controversy for many years, despite extensive research into every aspect of their nature. It is still not known whether they are of terrestrial or extraterrestrial origin. Some of the tektites contain gas bubbles and these have been shown to contain vapour at a very low atmospheric pressure, equivalent to the pressure at about 18 miles above the earth; argon isotopes in the vapour show typically terrestrial atmospheric ratios, suggesting that the glass must initially have solidified in the upper part of the Earth's atmosphere. In addition, they do not appear to have suffered any bombardment by cosmic rays as would be expected if they had spent any amount of time in space. These features, together with several other lines of evidence, seem to favour some kind of terrestrial origin – they might be fragments of fused rock ejected from the impact area of a large meteorite or comet for example – but there are many arguments against this, for nothing of a similar composition is known on Earth, and in addition it is thought to be aerodynamically impossible for such small bodies to be ejected from, and returned to, the Earth's surface. Other origins that have been suggested include volcanic eruptions on the Moon and lunar impact spallation throwing tektites far out from the Moon's surface. In any event, tektites represent either a most unusual form of extraterrestrial material reaching the Earth or are in some way formed on the Earth during or in consequence of a meteorite or comet impact with the atmosphere or surface of our planet.

Craters with associated meteorites are known from many places all over the world. They are obviously similar in origin to Barringer Crater in Arizona. At numerous other localities impact structures have been postulated, but, because the scars are old, erosion has removed all trace of the original ground surface and thus no associated meteorites or meteorite fragments can be found today to support

Table 7.4   Large impact structures on Earth

Name, locality, number of craters	Size (largest diameter)	Category	Approximate age (if known) in millions of years
Astrons (more than 20 already suggested) roughly circular features that may be the result of very ancient impacts by large planetisimals: distribution worldwide	100–2000 miles	Problematical	>3000 Myr.
Bushveld, South Africa (3)	each 70 miles	Probable	2100 Myr.
Puchezh-Katun, Gorkii, USSR	44 miles	Possible	?
Popigay, Taymirskayiy-Yakut, USSR	40 miles	Possible	?
Manicouagan, Quebec (Plate 25)	40 miles	Probable	220 Myr.
Labynkyr, Yakut, USSR	37 miles	Possible	?
Vredefort, South Africa	35 miles	Probable	2100 Myr.
Sudbury, Ontario	30 miles	Probable	1700 Myr.
Siljan, Sweden	28 miles	Probable	280 Myr.
Charlevoix, Quebec	22 miles	Probable	380 Myr.
Carswell Lake, Saskatchewan	20 miles	Probable	485 Myr.
Manson, Iowa	18 miles	Probable	135 Myr.
Steen River, Alberta	15 miles	Probable	95 Myr.
Ries, Germany	15 miles	Probable	15 Myr.
St Martin, Manitoba	15 miles	Probable	225 Myr.
Gosses Bluff, NT, Australia	14 miles	Probable	130 Myr.
West Clearwater Lake, Quebec	13 miles	Probable	285 Myr.
Lake Mistatin, Labrador	12 miles	Probable	200 Myr.
El'gygt-gyn, Chutotsk, USSR	11 miles	virtually certain	1 Myr.
Haughton Dome, NWT, Canada	11 miles	Possible	?
Strangways, NT, Australia	10 miles	Probable	280 Myr.
Zhamanshin, Aktyubinsk, USSR	9 miles	Possible	?
East Clearwater Lake, Quebec	9 miles	Probable	285 Myr.
Rochechouart, France	9 miles	Probable	165 Myr.
Wells Crater, Tennessee	9 miles	Probable	200 Myr.
Kilmichael, Mississippi	8 miles	Possible	?
Sierra Madera, Texas	8 miles	Probable	150 Myr.
Nicholson Lake, NWT, Canada	8 miles	Probable	350 Myr.
Dellen, Sweden	8 miles	Probable	280 Myr.
St Magnus Bay, Shetland	7 miles	Possible	?

Name, locality, number of craters	Size (largest diameter)	Category	Approximate age (if known) in millions of years
Bosuntwi, Ashanti, Ghana	7 miles	Probable	1.3 Myr.
Des Plaines, Illinois	6 miles	Possible	?
Lac Couture, Quebec	6 miles	Probable	350 Myr.
Lappajärvi, Finland	6 miles	Probable	280 Myr.
Janisjarvi, USSR	6 miles	Possible	?
Deep Bay, Saskatchewan	6 miles	Probable	100 Myr.
Eagle Butte, Alberta	6 miles	Possible	?
Elbow, Saskatchewan	5 miles	Possible	?
Wanapitei, Ontario	5 miles	Probable	280 Myr.
Middlesboro, Kentucky	4 miles	Probable	280 Myr.
Serpent Mound, Ohio	4 miles	Probable	280 Myr.
Hartney, Manitoba	4 miles	Possible	?
Decaturville, Missouri	4 miles	Probable	500 Myr.
Kentland, Indiana	4 miles	Probable	280 Myr.
Crooked Creek, Missouri	3 miles	Probable	320 Myr.
Köfels, Austria	3 miles	Probable	1.5 Myr.
Mien Lake, Sweden	3 miles	Probable	280 Myr.
Pilot Lake, NWT, Canada	3 miles	Possible	350 Myr.
Glasford, Illinois	3 miles	Possible	?
Upheaval Dome, Utah	3 miles	Possible	?
Brent, Ontario	3 miles	Probable	450 Myr.
Unnamed Lake, NWT, Canada	3 miles	Possible	?
Flynn Creek, Tennessee	2 miles	Probable	450 Myr.
New Quebec, Quebec	2 miles	Probable	0.15 Myr.
Steinheim, Germany	2 miles	Probable	15 Myr.
West Hawk Lake, Manitoba	2 miles	Probable	150 Myr.
Al Umchaimin, Iraq	2 miles	Possible	?
Gebel Dalma, Libya	2 miles	Possible	?
Jeptha Knob, Kentucky	2 miles	Possible	?
Skeleton Lake, Ontario	2 miles	Possible	?
Chubb Lake, Ungava, Ontario	2 miles	Possible	0.02 Myr.

Plus many lesser probable or possible impact craters and/or structures and numerous problematical structures of all sizes.

(NB Final resolution of disputed volcanic or impact origin for a ring structure can be made only by drilling deep boreholes in the crater.)

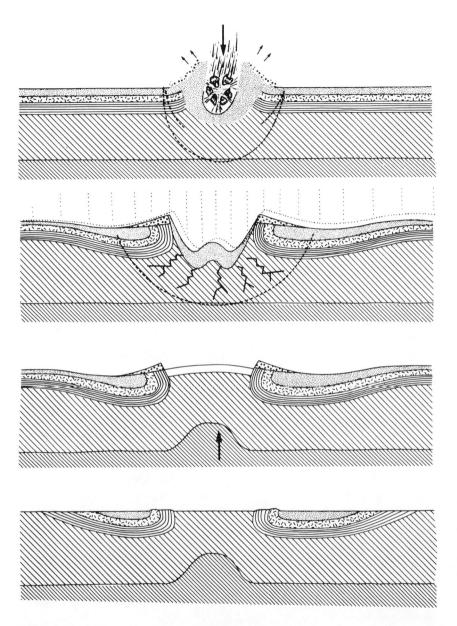

Fig. 33 Diagrammatic cross-sections of the Earth's Crust illustrating the development of the Vredefort astrobleme near Johannesburg in South Africa. The upper drawing shows the impact of a giant meteorite some 2100 million years ago. An enormous open crater forms and the meteorite itself is completely disrupted and liquefied by the exploding release of its kinetic energy. A powerful shock wave spreads out through the surrounding crustal rocks, causing brecciation, fracturing, the development of shatter cones, occasional veins of melted rock and growth of the rare high pressure silica minerals *coesite* and *stishtovite*. In the next drawing, layers of sedimentary rocks overlying the gneissic basement are seen to have been thrust upwards and overturned away from the crater. After the initial impact energy has been dissipated, the Crust recoils to form a central uplifted dome in the crater. A thick collar of

the impact hypothesis. One of the largest circular scars of questionable origin is the huge 2100 million year old structure known as the Vredefort Dome or Ring in South Africa. One of a group of four astroblemes, but, unlike the others, not overlapped by its neighbours, it has been examined in great detail by geologists. The centre of the Vredefort structure is occupied by a plug of brecciated, fused and reworked ancient granite 25 miles in diameter, surrounded by a 50,000 feet thick girdle of upturned and overtilted Precambrian sediments, highly metamorphosed in a zone extending a remarkable 3.5 miles from the margin of the central granite plug. The old granite is thought to have been uplifted to the surface from a depth of at least 9 miles and it has been suggested that the whole Vredefort Dome represents the substructure of a gigantic, 35 mile-wide, 10 mile-deep, meteorite explosion crater, into the centre of which a huge plug of remobilized basement granite was thrust up from depth later, obliterating all trace of the original crater. It has been calculated that to cause such catastrophic damage to the Crust, an explosive force of more than 1.5 million megatons of TNT would be required – resulting from collision with an asteroid (or planetisimal) over a mile in diameter. In one variant of the hypothesis it is assumed that the colliding body passed right through the Crust and lodged in the Earth's Mantle. The central granite and the sedimentary collar of Vredefort are extensively veined by anastomizing veins of black *pseudotachylite*, a fine-grained rock of originally glassy texture, which is rarely found elsewhere in the Crust, and nowhere on such an extensive scale as here. This is thought to have been formed by fusion of materials in the Crust under the extreme and sudden pressure of the shock wave caused by the impact. *Shatter cones*, rock structures thought to be a significant criterion for meteorite impact structures, are found in great abundance in all the rocks of the Vredefort Ring. Other evidence of intense pressure is the profusion of shattered and crushed rocks and numerous faults.

Barringer Crater, Arizona and the Vredefort Dome in South Africa are two contrasted examples of the structures known as astroblemes – the geological scars left by the ancient impact of cosmic

---

ejected breccia forms the crater rim. Finer debris settles gradually over a wide area. Everywhere on Earth suffers some fallout from the impact. The third drawing shows the eventual disappearance of the deep open pit as isostatic equilibrium is regained by a slow upwelling of the Crust beneath it. The final drawing shows Vredefort today. Erosion over thousands of millions of years has bevelled off the higher parts of the structure, leaving no direct evidence of the original crater or of the surface layer of ejected debris, but the great upturned ring of sediments is still very clearly seen surrounding a central core of upwelled gneiss. In addition, the shatter cones and high pressure minerals produced by the passage of the shock waves can still be found in the rocks.

objects. One is definitely a large meteorite, the other possibly a much more massive missile, the size of a small planet or asteroid. There is, of course, considerable controversy concerning the origin of the largest astroblemes, since actual meteorite remains have only been found associated with a small number of the structures. This is not as surprising as it may seem, however, because the larger masses of cosmic debris, those that would cause huge scars if in collision with the Earth, suffer little braking effect by the atmosphere and can arrive at extremely high velocities; such high velocities might lead to instantaneous explosion and vaporization of much of the body on impact. The presence of two particular forms of silica, *coesite* and *stishtovite*, points to formation during an intense meteorite impact event; these are minerals which are only formed under extremely high pressures and do not occur naturally in the Crust under any other circumstances. In addition, there are a number of other features, common to craters of all sizes and ages, which show the effects of shock metamorphism on the country rocks. Shatter cones are conical fracture surfaces, covered in striations which fan out from the apex of the cone. They are usually found at their best in limestones, but may also be developed in many other rock types, and may vary in length from ½ inch long to several feet. These structures are only found in rocks within presumed meteorite impact craters, and have been found associated with at least 20 major astroblemes. They are caused by the sudden application of intense overpressure to the host rock; the rock instantly contracts and expands in reaction to the shock waves generated by the impact, and fractures in a characteristic pattern of stacked shatter cones, towards the point of impact. Shatter-cone formation requires overpressures in the range 300,000 to 1,200,000 lb/square inch, far higher than any that could be produced by explosive volcanic activity, for example. The rocks within impact craters commonly show intense *brecciation*; that is, they are broken up into angular fragments, which vary in size from huge boulders to dust-sized particles and occur as a chaotic mass infilling the crater and ejected from it to form a mantle on the surrounding ground. In some places the breccias are veined and cemented together by glassy melt rocks, formed by the total melting of some of the rocks at the impact site. In some craters the deepest part may be infilled with a pool of impact melt rock. Glass 'bombs', similar in appearance to lava bombs produced during volcanic eruptions, may also occur in the shock breccias found in and around the craters.

The initial identification of an impact crater is often a result of its circular outline when seen on a geological map or by aerial survey.

Although usually distinctly circular near the centre, the margins of large craters may be obscured by pre-existing structures in the country rock covered by later sediments or destroyed by subsequent erosion. One of the largest structures that is thought to be an undoubted astrobleme, the El'gygt-gyn crater in northern Siberia, was identified by aerial imagery from satellite photographs. This crater is some 11 miles in diameter: its rim is marked by a ring of mountains highest in the west, where they rise 1350 feet above the level of the lake which now occupies the central part of the crater. Such obviously raised rims are only found in relatively recent, simple craters; in older or more complex craters the rim has either been destroyed by erosion or has subsided to form a subdued uplift or is represented by a peripheral trough. Complex craters over one mile in diameter also commonly show an area of central uplift, formed by the elastic rebound of the rocks immediately below the point of impact. Very similar structures have been produced artificially by experiments in which large spheres of TNT were detonated at ground level, and it seems from this evidence that some large, complex astroblemes might be the result of impact by cometary bodies, which, while they would generate intense shock waves at the surface would be entirely volatilized on impact.

Geophysical features associated with large impact craters include a typically negative gravity anomaly over the crater, indicating the presence of rocks of low density beneath the surface; this is most clearly developed in craters of moderate size. The crater rocks also usually show lower seismic velocities than the country rocks as a result of their chaotic structures. Nevertheless, where a large heavy missile has been completely swallowed by the Earth, a positive gravity anomaly and higher seismic velocities might be expected.

Over 50 large astroblemes can now be identified with reasonable certainty on the continental land surface of the Earth; a much longer list of possible and problematical impact structures could be drawn up, to include features of the Earth's surface with diameters measurable in hundreds of miles down to quite small structures, some of which show evidence of shock metamorphism but none of which have associated meteorite remains. Any list of astroblemes prepared for the Earth, however, will of necessity be an underestimate of the true extent of extraterrestrial bombardment of our planet, because we have no record of any impact that has occurred in the oceans and the evidence for many that did hit the continents lies buried beneath a cover of younger sedimentary rocks. Two probable impact structures which occur in Europe are the Ries and Steinheim Basins, 22 miles apart in southern Germany. The Ries Basin is a circular depression, 15 miles in diameter with a surrounding rim of hills standing some 300 feet above the central plain; it is about 15 million years old. Although much of the central part of the crater is covered by later sediments deposited in a lake, a thick layer of *suevite*, a breccia showing strong shock features, occurs blanketing the central part of the crater where it is up to 1200 feet thick. It also occurs as a 'fall-out' deposit in the area surrounding the crater, extending to a distance of 10 miles from the crater rim. This rock consists of fragments of crystalline country rock, mineral and glass fragments and glass 'bombs' in a very fine-grained, dusty matrix. The high pressure silica minerals, coesite and stishtovite, occur in the shocked rock fragments. The suevite mass appears to have been deposited as a hot pyroclast flow density current, flowing out from an eruptive column ejected more or less vertically from the centre of the crater. The glass 'bombs' represent material which suffered total melting during the impact and was erupted in this column. There are also extensive deposits of a more normal, unshocked type of breccia; this contains blocks up to half a mile in diameter near the rim of the crater, giving some idea of the force of the impact. Geophysical surveys of the crater have shown negative gravity and magnetic anomalies over the crater and a complicated seismic picture at depth in the crater. The nearby Steinheim Basin, thought to be of the same age as the Ries Basin, is a much smaller depression, 2 miles in diameter, lying some 300 feet below the surrounding plain. This crater contains a central uplifted area of chaotic, fractured and brecciated sediments, with many fracture cones. It is possible that both impacts occurred at the same time, produced by a single event in which several impacting bodies hit the Earth.

All the craters which are *accepted* as having been formed by certain meteorite impact are of geologically recent age – not more than 10 million years old. The *probable* and *possible* impact sites, such as those described above, and many older structures, some of which only show the features of shock deformation, have all been affected to a greater or lesser extent by erosion, subsequent deformation and cover by younger sedimentary rocks. Thus, even if they were once there, it is not surprising that no actual meteorite remains are found at these older sites. Some geologists have suggested that the largest arcuate features of the Earth's Crust may be related to ancient impact wounds. In their opinion the western bulge of Africa and various arcuate mountain ranges and trench features such as the Aleutian Arc and the East Indies may all have fundamental controls related to impact scars. Definitive evidence is difficult to obtain in such instances: obviously the older an impact structure is, the more will its original form have been eroded and masked, and the more problematic will be its origin. It is quite likely, however, that some of the oldest rocks on Earth, the Precambrian greenstone belts of the ancient continental shield areas, may owe their origin to huge impact events which took place early in the history of the Earth, around 4000 million years ago. The scale of these ancient structures – called *astrons* – means that the impacting bodies would have had to be some 200–300 miles in diameter; on impact, such a huge body would have been completely swallowed by the Earth and would have triggered intense melting within the Crust and Upper Mantle, leading to voluminous volcanic activity in the crater. The result would have been enormous impact scars on Earth similar in appearance to the maria on the Moon and, like the lunar maria, infilled with lava flood volcanics exceptionally rich in ferromagnesian minerals. On the mobile Earth, however, such large 'maria' would soon be obliterated by the folding or overthrusting of the surrounding primitive Crust over the crater volcanics, transforming a circular lava-filled surface impact structure into an elongated infold of metamorphosed volcanics. Certain distinctive features of the rocks in the Precambrian greenstone belts of Swaziland, Rhodesia, Western Australia, Greenland and elsewhere have led some geologists to suggest this form of extraterrestrial origin. It is interesting that these events, if they occurred, would have been at about the same time as the well-documented large impact events which formed the maria on the Moon.

In order to put all of the impact possibilities on Earth into perspective, it is necessary to take a closer look at the impact structures to be seen on the Moon and on the nearby planets of the Solar System in the light of the knowledge recently acquired from the American and Russian space exploration programmes. The lunar surface has been intensely observed since the first detailed investigation by the Italian astronomer Galileo in the seventeenth century, using the newly-invented astronomical telescope. The idea that the craters of the Moon were formed by meteorite impact was first suggested in the early nineteenth century and this view was extensively developed in 1892 by the American geologist Gilbert. During the first half of the twentieth century it was generally thought that all the lunar craters were produced by the continuing impact of natural rocky missiles varying in size from small meteorites to small planets or asteroids. However, with the publication of the first close-up pictures of the

Moon, taken in 1964 by Ranger 7, it became apparent that the ideas of a dissenting group of geologists – who maintained that volcanic processes also played a major part in forming the surface features of the Moon – were substantially correct. Further support for this view-point was soon forthcoming from the many thousands of pictures taken by the succeeding Ranger, Surveyor, Lunar Orbiter and early Apollo missions to the Moon (Plate 24). With the lunar landings of Apollo 11, 12, 14, 15, 16 and 17, between 1969 and 1972, came the extensive sampling of many different types of lunar rock and since they arrived on Earth these rocks have been the subject of intensive research, discussion and controversy. Although many of the details are still incomplete, we now have a vastly increased knowledge of the internal structure of the Moon, the composition of its crust, the processes which formed this crust and the sequence of events during the evolution of major areas of its surface. Apart from confirmation of the hypothesis that many of the lunar surface features are volcanic in origin, another surprising fact to emerge from laboratory investiga-tion of lunar rocks has been the extreme age of almost all the rocks found at the surface. Few of the great variety of rocks sampled have yielded ages younger than 3000 million years, the majority are be-tween 3500 and 4000 million years old – as old as the oldest rocks ever found on Earth – and a number are even older than 4000 million years. This immediately implies that the major surface geological activity on the Moon, whether it be volcanic activity or impact events, was very largely finished by 3000 million years ago and that since that time the Moon has remained a barren, geologically quiet planet, disturbed only by an incessant rain of smaller meteorites, a few larger collisions, some minor volcanic eruptions and a few major earth-quakes. By contrast, the same period on Earth has seen the develop-ment of the geologically complex terrestrial Crust as we know it today, the product, as we have seen, of an incredible history of violence and mobility.

When we look at the surface of the Moon facing the Earth, a number of major features are visible even with the naked eye. The dark areas, or *maria*, are huge, flat depressions, once thought to be oceans or the remains of ocean basins, but now known to be areas of extensive lava flows. The lighter surrounding areas, the lunar high-lands, are highly-cratered areas of lighter coloured and brecciated rocks; these breccias are largely composed of *anorthosite*, an igneous rock rich in plagioclase feldspar (calcium-aluminium silicate) that originally crystallized at some depth below the surface. On the far side of the Moon, never seen from the Earth, there are no large maria as such; almost the whole surface has the highly-cratered appearance of the lunar highlands of the near side. Until the recovery of rocks for age determination by the manned Apollo missions, it has been assumed that the craters of the Moon were the visible expression of the continual bombardment of the lunar surface from its time of formation to the present. We now know, however, that the vast majority of the craters on the lunar highland surfaces were formed during a period of intense bombardment in the early history of the Moon, between 4600 and 4000 million years ago, and that this early period culminated in a series of major impacts leading to the formation of the mare basins.

The outer crust of the Moon is thought to have been completely consolidated and solidified in roughly its present form by about 4400 million years ago. During the latter part of this process, the Moon was constantly bombarded by violent storms of debris, the remnants of the primordial cloud from which the planets of the Solar System were born. Even after a recognizable lunar crust had formed, this bombardment continued, completely saturating the outer crustal layers we now see as the lunar highland areas with repeatedly overlapping impact craters, 25 to 55 miles in diameter or possibly much larger. This intense bombardment affected the crustal material to a depth of about 20 miles. Soon after 4000 million years ago, the large basins of the lunar maria were excavated by collision with a small number of enormous impacting bodies; at the time of impact, shock-melted lavas were probably generated in the mare basins, but few remnants of these rocks can be found on the surface, as the mare depressions are now completely filled by thick accumulations of younger basaltic lavas poured out from fissure volcanoes between 3800 and 3000 million years ago. During the impact events which formed the maria, huge amounts of pulverized crustal material were flung out and transported in a manner similar to terrestrial pyroclast flows to be deposited as a thick blanket of debris over vast areas, often hundreds of miles from the original impact sites. Since the end of the main period of volcanic activity there have been occasionally fairly large meteorite impacts forming the young, prominent 'rayed' craters such as Copernicus, formed about 900 million years ago and Tycho, around 100 million years ago. The most recent crater examined by Apollo 17 astronauts was found to be only 2 million years old, so it is certain that crater formation has continued until today and that further cratering will occur in the future. A recent search of medieval records has brought to light what is probably the only recorded observation of an actual meteorite impact on the Moon, written by

Gervase of Canterbury in England in AD 1178. He described it as follows:

> . . . a flaming torch sprang up, spewing out over a considerable distance fire, hot coals and sparks. Meanwhile the body of the Moon which was below writhed, as it were in anxiety . . . and throbbed like a wounded snake. Afterwards it resumed its proper state. This phenomenon was repeated a dozen times or more, the flames assuming various twisted shapes at random and then returning to normal. Then after these transformations, the Moon from horn to horn, that is along its whole length, took on a blackish appearance.

The chances of such an observation from Earth over the last 3000 years of semi-recorded history is, at best, about 1 in 2000, even assuming continuous observation of the Moon; therefore such a sighting is an extremely fortuitous event. A very young, fresh crater on the surface of the Moon, *Giordano Bruno*, has been identified as the probable result of this impact event.

The cratered surface of the Moon has been known in some detail for hundreds of years, but it came as a surprise to observers on Earth when the first photographs of Mars and its satellites, together with those of Mercury and Venus, showed that these planets also had heavily cratered surfaces. Mariner and Viking photographs of Mars have shown that, like the Moon, much of the Martian surface has supported substantial volcanic activity; a number of huge craters on the surface are calderas of volcanic origin. The largest of these, Olympus Mons, is probably the biggest volcano yet identified in the Solar System, with a base diameter of over 1000 miles. However, as in the case of the Moon, it is obvious that most of the craters on the surface of Mars do not result from volcanic activity but from meteorite impact. Viking Orbiter pictures of Phobos, one of the satellites of Mars, have shown that it too has a heavily cratered surface. Its tiny dimensions (10×12 miles) mean that it could never have had a hot enough interior to support volcanic activity: so in this case all of the craters must be the result of meteorite impact. Mariner photographs of Mercury show the same heavily-cratered appearance as the lunar highlands. The craters seen on the surface of Mercury are generally less than about 100 miles in diameter and some volcanic features can also be picked out, so again it seems likely that at least some of the craters are volcanic in origin. Radar photographs of Venus, which is similar in mass to the Earth and also has similar, raised, 'continental' regions on its surface, have shown that it, too, has a cratered surface; the largest crater has a diameter of about 90 miles with a 1500 feet high rim. In the case of Venus, the degree of surface volcanism is completely unknown, but it seems reasonable to assume that, as in the case of the Moon, Mars and Mercury, its craters are of mixed origin.

The similarity between the larger meteorite craters on the Moon and the astroblemes found on the surface of the Earth can be seen in a number of features: the circular or formerly circular layout of the craters, often with a raised central portion formed by the crustal rocks rebounding under the impact of a large body; the shock metamorphism features, such as impact-melted rocks in the craters; the ejecta or breccia thrown from the crater, which on the Moon often forms characteristic 'rays', radiating from the crater, with lines of subsidiary craters formed by large fragments of breccia and by many other phenomena. Until the extreme age of most of the lunar craters was discovered, it had been assumed that the much denser pattern of impact structures seen on the Moon was largely the result of its lack of an atmosphere in which most meteorites would burn out before hitting the surface. It now appears that the rate of bombardment of the Moon has changed markedly with time and that most of the lunar craters can be ascribed to an early, intense phase of asteroid and meteorite bombardment in the distant past, when both the Moon and the Earth were in the process of formation. While the lack of atmosphere on the Moon has meant that it has had a higher rate of bombardment by small meteorites than that experienced below the Earth's thick atmosphere, this fact alone is not the prime cause of the difference in observed impact patterns between the Earth and its satellite. We can now see little or no evidence of what was happening at the surface of the Earth during the earliest periods in its history. Rocks of such antiquity are found only in a few isolated areas in the Crust, and have been often so changed by later geological activity that any impact features they may have shown have long since been disguised or obliterated. While the Moon has remained largely frozen and static for the last 3000 million years, the Earth, by comparison, has been a boiling, violently active planet of constant geological change and upheaval. In the record of the last 3800 million years that we can discern in the Earth's geological column, there are few, if any, clearly recorded occurrences of meteorite impact structures or meteorite remains. Among the structures we can recognize on the present surface of the Earth, only a comparatively few can be assigned to an asteroid or meteorite impact origin with absolute certainty, and even if all of the more questionable astroblemes are included in our survey, it is apparent that the impact of very large, crater-forming meteorites or asteroids is now a comparatively rare event on Earth. Careful counting of the craters on the Moon has revealed that earlier than 4000 million years ago the rate at which the Moon was being cratered was hundreds or even thousands of times greater than the rate at which it has been cratered recently. Moreover, it is clear that the rate declined rapidly from that early time until about 2000 million years ago when it reached something like the current level. The rate since that time has been relatively constant. It is now believed that the other inner planets and satellites, including Earth, experienced a similar impact sequence. The spectrum of sizes of the meteorites and asteroids that formed the craters on the Moon and the inner planets is essentially the same as the range in size of the space debris seen today in the asteroid belts. The meteorites that fall on the Earth belong to this spectrum: it is a type characteristic of fragments created when large masses of rock collide at high speed.

To summarize current knowledge: the Solar family of planets appears to have formed some 4600 million years ago when the matter in the cloud of gas and dust surrounding the primordial Sun gradually collected under gravitational attraction into larger solid objects. As the protoplanets accumulated they must have been intensely bombarded by the remaining objects both large and small. The final stage of the initial sweeping up of the debris left over from the primordial nebula might have taken only a few million years and it would appear that during this early period sufficient heat was generated for molten rock (magmas) to form and for very considerable igneous activity to occur on and near to the surfaces of the protoplanets, resulting in the development of thick crystalline crustal layers. Thus the oldest rocks found on the surfaces of the inner planetary bodies are all about 4600–4400 million years old. These rocks are, however, found only as fragments because this was also still a period of continued intense bombardment during which the primitive crusts were being quickly pulverized in overlapping impact craters. About 4000 million years ago one of the smaller protoplanets was fragmented, possibly by tidal forces. Numerous collisions between the ensuing fragments followed. The debris derived from this disrupted protoplanet is contained in the asteroid belt today and forms much of the reservoir of small inter-planetary objects and meteorites that have been bombarding the other inner planets in more recent times. Age determinations on meteorites reveal that their parent protoplanet formed between 4600 and 4400 million years ago. The largest bodies formed in its disrup-tion were a few hundred miles across. Some collided with the Moon producing the lunar maria and some with the Earth. If all the clearly recognizable craters which are seen on the inner planets are assumed to have been formed by missiles which came from the asteroid belt, the original number produced must have been about twice that still available today. Volcanism was an important feature of the larger terrestrial planets and the Moon 3000–4000 million years ago. Only on the Earth, however, was the internal production of radioactive heat sufficient to cause intensive crustal mobility, to make volcanism a permanent feature, leading to the growth of a hydrosphere and a relatively dense atmosphere and to provide the driving force necessary to sustain a complicated geological history of repeated uplift, mountain building, erosion, sedimentation, magmatism and metamorphism.

What, then, is the risk to man from extraterrestrial bombardment? Although the evidence is not preserved explicitly on the Earth as it is on the Moon, Mars and Mercury, modern scientific observation of actual meteorite falls, the recognition and cataloguing of a number of fairly recent meteorite impact craters and the study of the numerous ancient astroblemes distributed over the surface of our planet show conclusively that the Earth has been and will continue to be subjected to the same kind of damaging bombardment from space that has been experienced by the Moon and the other inner planets. Thus, the likelihood of Earth colliding with a planetisimal or asteroid as large as those which produced Copernicus, Aristarchus and Tycho craters on the Moon (900, 600 and 100 million years ago respectively) in the geologically near future must be quite high. Indeed, during the same period that saw the impact events mentioned above on the Moon a number of similar impacts must have occurred on Earth. We know of only a few of them – those which struck land areas and which have been recognized as such. Deep Bay crater at the south end of Reindeer Lake in Saskatchewan, for instance, is over 5 miles across and is undoubtedly an astrobleme that represents an impact on Earth of approximately the same age as Tycho on the Moon. At its worst, if an impact event of the magnitude of the Tycho, Aristarchus or Copernicus events on the Moon occurred on Earth today, it might well cause the complete extinction of man. At its best, its effects would be catastrophic on an unprecedented and almost unimaginable scale.

The extensive, relatively flat plains formed by the consolidation of basalt lavas in the lunar maria enable us to examine the results of a large meteorite impact on a flat land surface. It can be seen that when meteorites a mile or more in diameter have struck the smooth lava surface of lunar maria, the impacts have caused enormous explosions, each excavating a huge crater and lifting the layers of lunar lavas to form an upturned crater rim often over a mile high. At the same time high-speed jets of material shot outwards from the explosion centre and fell on the surface as long streamers of dust and glass beads, seen as bright radial rays visible from Earth. Low-speed ejected fragments fell back as a thick blanket of hummocky terrain surrounding the crater for several times its diameter. The rain of material was heaviest on the crater rim, thus adding considerable height to the uplifted rock layers. Many isolated blocks of debris thrown high above the impact point fell back nearby, simulating meteorites and themselves forming further small craters. On the Moon there is no atmosphere in the form that we know it on Earth, but even so, around some of the larger impact areas on the Moon debris has moved across the surface in pyroclast flow form as a result of the fluidization of debris by gases released during the instantaneous fusion of rocks in the impact area. It is very likely that the area surrounded a similar large impact crater on Earth would be inundated by much more extensive pyroclast flow. An added catastrophic effect in the area surrounding an Earth impact crater, of course, would be the blast wave travelling through the atmosphere. In addition, not only would the primary seismic shock waves travelling out from the impact area be themselves destructive but they would trigger many violent earthquakes in distant areas where stresses were already at a high level in the Earth's Crust and cause very considerable seismic sea waves on all of the larger bodies of water distributed over the Earth's Crust.

Even if a large missile approaching Earth from space were to miss the heavily-populated areas within the continents or the equally densely-populated ocean shores and struck, say, in the middle of the Pacific Ocean, the results would still be catastrophic. The resulting shock wave would destroy marine life over a wide area, while the resulting king-sized tsunami would be possibly between ½ mile and 1 mile high, and would sweep away from the impact area causing widespread havoc among terrestrial life of all forms in adjacent low-lying coastal areas. The steam, water vapour and dust generated by the impact would rise into the outer atmosphere and influence the world's weather for a considerable period. Indeed it is not inconceivable that the impact and swallowing of a sufficiently large extraterrestrial body could induce the onset of an ice age or severely modify the Earth's magnetic field. The combined effects of such an event, causing instant destruction of life over a wide area, the modification of climate and the Earth's magnetic field could indeed be disastrous in every way. It is not impossible that the sudden demise of certain species in the past – for instance, the dinosaurs – may be attributable to such a cosmic catastrophe.

Some idea of the size of the immediate area of complete devastation that would be caused by a huge impact event occurring on land can be obtained from the superimposition of Copernicus crater (56 miles in diameter) onto a map of any one of the continents of the Earth. The energy released in such an impact might be between 10^{10}–10^{35} ergs (equivalent to many hundreds of million megaton nuclear devices exploding at one spot). The magnitude of the shock waves generated by such an impact would be enormous, with a maximum measurable in megabars (millions of times greater than atmospheric pressure), a pressure only normally encountered in the Earth's Core. A pressure wave of such extreme intensity would sweep out from the impact point, melting, deforming and welding the rock particles of the Crust together and inducing phase changes in its constituent minerals. Of course, as the shock waves spread out they would progressively lose energy until a point was reached where they were no longer able to produce plastic deformation within the rock but would still have the energy to produce brittle fracturing. Away from the actual crater with its surrounding ring of drop faulted country, the seismic shock waves would traverse the entire globe, causing and triggering earthquakes of decreasing intensity away from the impact point. Clearly, the immediate loss of life and complete destruction of property as a result of the direct impact explosion would be of almost continental proportions. Loss of life elsewhere on Earth, far from the area completely wiped out by the direct explosion, from the secondary effects of Earth-encircling tsunamis, shock-induced earthquakes and volcanic eruptions would also be enormous, as cities and building works of all sorts would be destroyed or badly damaged. Indirect loss of life due to subsequent famine and pestilence, short- and long-term climatic and radiation level changes might eventually result in a much higher death toll figure than that of the immediate loss of life and, indeed, might in certain circumstances be so great as to cause the eventual extinction of mankind as a viable species. Even if a less gloomy view is taken, the effects of a large impact event would be cataclysmic to man, his works and his fragile political arrangements. It would certainly change life on Earth as we know it out of all recognition. Smaller impact events would have effects equally disastrous but of lesser extent. The impact of a small comet or asteroid, for example, could have as devastating an effect on a populated area as we anticipated might result from a man-made thermonuclear holocaust.

Let us assume that a missile from space, about the same size as one of the many large bodies which we know to have hit the Earth and Moon during the last 600 million years, strikes the United States of America in the vicinity of Columbus, Ohio. What would the likely effects be of this catastrophe? Clearly all life would be extinguished in the impact area and for a radius of many hundreds of miles as a result of the direct blast of the impact explosion. The effects of air blast and ground shock waves combined with the incredible rain of flying debris ejected from the crater and the surge of incandescent pyroclast flows across the ground surface would certainly kill people throughout the eastern half of North America. Secondary earthquakes, seiches and tsunamis would kill many more: in all, perhaps more than 30 per cent of the North American population might die immediately. This would not be the end of the death toll: many more would die as a result of the ensuing collapse of civilized amenities and from the effects of intense climatic and geographical disruption. Famine, disease and plague would sweep across the land and eventually very few would survive. It would be the end of North American civilization as we know it today.

In the place of Columbus and the surrounding areas there would be a huge impact crater, say some 50 miles across, with upturned rock debris-covered walls several miles high. For hundreds of miles around this crater the original ground surface would be covered by a thick gently outward-sloping fan of airborne debris, pock-marked where giant blocks and boulders had fallen from the sky and cut by faults, horsts and graben caused by subsidence and upheaval consequent upon the rebound of the land surface after the passage of the shock waves. Cities as far away as Chicago, St Louis, Memphis, Knoxville and Cleveland would have been completely destroyed by the combined effects of the ground and air shock waves and would be partially or completely buried by the fan of airborne debris. In certain apparently random directions, large quantities of debris would have been flung out as great jets and enormous incandescent fluidized streams of rock fragments would have rushed outwards from the impact crater, crossing the ground at speeds of more than 100 miles an hour. Some of these distinct 'rays' of white-hot debris would have reached out as far as the coasts of the continent, killing, maiming and burying everything in their path. Immediately after the impact event these cooling and consolidating 'rays' of debris would be like great barriers dividing what was left of the eastern half of the North American continent into a number of separate sectors. In the intervening sectors, even far away from the central crater and the debris-buried areas, damage would be considerable. Any region close to a lake or river or on the coast would have suffered from the effects of seismic disturbance of the water, causing the creation of great waves and tsunamis; on coasts, the sudden withdrawal of the sea would have been followed by its catastrophic return as a gigantic wall of water sweeping onto the land. The damage and loss of life in places like Toronto, New York and San Francisco from these causes alone would have been terrible. Elsewhere, earthquakes would have rent the ground, mountains would have collapsed, rivers diverted, dams

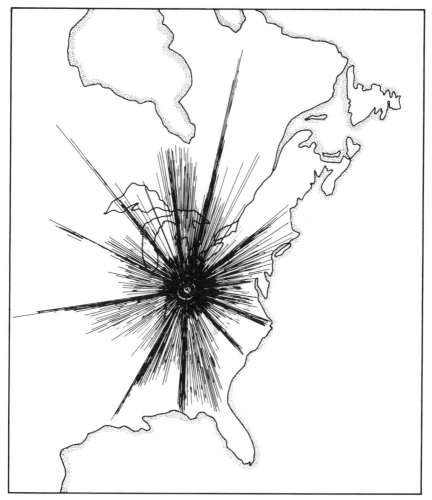

Fig. 34   The effect of a small astronomical body no larger in size than several which we know have collided with the Earth or the Moon within the last 600 million years striking the Earth at Columbus, Ohio would be more devastating than any of the bizarre disasters visualized for this city by James Thurber. Columbus would be replaced by a huge crater, some 50 miles across, rays of incandescent debris would spurt out halfway across the continent and some 30 per cent of the population of North America would die. Civilized life would cease east of the Rockies.

failed, oil wells blown and every other conceivable kind of disaster would have occurred. In the aftermath the weather would most probably have gone mad, with gigantic hurricanes and thunderstorms to add to the looming horror of the dark dust-filled skies. The loss of life in North America after an impact of this size in the middle of that continent would, of course, be worse than elsewhere in the world, but loss of life elsewhere would also be of catastrophic proportions, for gigantic tsunamis would have swept out from the American continent to engulf the oceanic islands and the lowlands of the land areas surrounding the Atlantic and Pacific Oceans. These giant sea waves, even after crossing thousands of miles of ocean, might well be hundreds of feet in height and the loss of life from this cause alone would certainly run into many additional millions. Other world-wide secondary effects, such as earthquakes, climatic disasters, pestilence and disease would certainly account for many more deaths. The political repercussions, especially in the event of man's selfishness overcoming his compassion, might eventually lead to an even greater death toll. Certainly things would never be the same again.

It is clear from the evidence we have reviewed in this chapter that man is in constant, if unlikely, danger of bombardment from space. Occasionally a single individual may be injured or property slightly damaged by one of the frequent falls of small meteorites that strike the surface of our planet at a rate of about three or four every day. Fortunately the larger meteorites fall at progressively less frequent intervals and thus the actual individual risk of death from meteorite impact is very low indeed, even if we include objects as large as those which made Barringer Crater in Arizona and Wolf Creek Crater in Australia. It follows that the mathematical probability of a collision between the Earth and a large extraterrestrial body, such as an asteroid or planetisimal several miles across, occurring within any one person's lifetime is extremely remote. Nevertheless, the possibility does exist and our review has shown that collisions of this kind occurred frequently in the geological past and will undoubtedly occur again in the future. If the next major impact of an asteroid or large comet onto the Earth is in the far distant future, perhaps beyond the lifespan of man, so much the better. But the fact remains that although it has a low probability, such an impact could occur at any time. Would it not, therefore, be prudent for some international body to be formed to examine and continuously review the problem, to co-ordinate our space watch efforts (at present piecemeal and often carried on under a cloak of secrecy), so that we could be sure of ample warning of any object approaching the Earth's orbit and, finally, but most importatly, in the event of a dangerous object being identified in space, to become the world-wide information and control centre of our efforts to avoid or mitigate the disaster? If the only task given to this international body were to be the prevention of mistaken nuclear retaliation after an unexpected natural impact, then it would still be immensely worthwhile.

8   Catastrophe or Survival?

> Well I know who'll take the credit – all
> the clever chaps that followed –
> Came, a dozen men together – never knew
> my desert fears;
> Tracked me by the camp I'd quitted, used
> the water holes I'd hollowed.
> They'll go back and do the talking. *They'll*
> be called the Pioneers.
>
> R. Kipling, *The Explorer*

Mankind has survived many great natural disasters. We have direct and often detailed knowledge of those that have occurred in the recent past, and an incomplete and less accurate record covering the previous two or three thousand years. Locally, after the worst of the historical catastrophes – such as after the 1556 earthquake in Shensi Province, China – the surviving population appears to have lost the will to live, but this response is temporary, and it is abundantly clear that the human species has great resilience and an incredible power to recover from disaster. Surprisingly, however, considering the insight and ingenuity shown by many individuals in their struggle to survive, mankind as a whole has been very slow to mobilize and co-ordinate any kind of natural disaster prediction, control and relief effort, except either locally or on an entirely *ad hoc* basis and, most frequently, *after the event*, when it is too late. In many human cultures there appears, in fact, to have been, and even still is, a blind and fatalistic attitude towards natural disaster, resulting in warning signs being virtually ignored and the inevitable catastrophes, when they come, being regarded as due to the 'anger of the gods', and thus totally beyond man's power to control. Even today, with our greatly expanded knowledge of the natural world, we continue to make what are often no more than puny and unco-ordinated efforts to mitigate the predictably calamitous effects of quite simple and well-understood processes of nature. Every year, tens of thousands of people are killed because the lessons of the past are totally ignored.

Geological studies reveal that, during its long history, our planet has suffered an endless succession of catastrophic natural events, many of which were several orders of magnitude greater and more damaging than those of recent or historical times. We know for certain that, in the future, innumerable potentially disastrous natural events, both large and small, are inevitable. This is a fact of life on Earth: a daunting prospect in which the death of millions of innocent victims can be foreseen. Nevertheless, it is the firmly held contention of the authors of this book that, while the endless succession of natural events is inevitable, most of the deaths are not. If the knowledge, resources and ingenuity of mankind were to be properly mobilized in a satisfactory national and international hierarchy devoted to disaster prediction, control and relief, a very considerable saving of life would be made. We see this as the great challenge of the future – conservation of man himself being as important as conservation of his environment.

In our present state of local, unco-ordinated semi-preparation for natural disaster, the relevant civil authorities often find it very difficult to select the unequivocal, expert scientific advice that they so urgently require from the plethora of conflicting advice with which they are inundated as soon as any great disaster seems imminent. This unfortunate and potentially dangerous situation is nowhere better illustrated than by our response to volcanic disaster, where, in some instances, as much suffering has been caused by panic evacuation as would have resulted from the feared disaster if it had occurred. It is mankind's lack of preparation and forward planning that is to blame: without a recognized World Disaster Authority to turn to for advice, how can local administrators be expected to distinguish between the expertise of one scientist who calls himself a 'volcanologist' and another, perhaps a well-known face from television, who is labelled a 'petrologist' or a 'geophysicist'? Again, the advice received from some of these so-called 'experts' will not be disinterested: scientists crave fame as strongly as, or sometimes even more strongly than, people in other professions.

In the past, there have been many instances where scientists who were totally unsuitable to advise on disaster procedure gave opinions which were based on inexperience and ignorance or, in some cases, deliberate falsification of the evidence. In 1972, the Soufrière of St Vincent, West Indies, started to erupt and produced a lava dome which extruded into the crater lake. In spite of a statement from a volcanologist on St Vincent who was actually examining the situation, indicating that there was nothing to fear, the entire population in the north part of the island was evacuated, at great expense, on the receipt of an isolated opinion from an 'expert' thousands of miles away who had made his observations and deductions from looking at local television coverage and earth-satellite imagery. Again, in 1973, an eruption on the island of Heimaey all but buried the thriving fishing port of Vestmannaeyjar. Initially the Icelandic authorities had wisely decided to use high explosives to breach the lava stream that was threatening their town and divert the flow into the Atlantic Ocean to the south. A foreign 'expert' with very little experience of volcanology persuaded the authorities not to do this on the grounds that it would expose vast quantities of molten rock to equally vast quantities of sea water, and that this would set up a chain reaction of explosions culminating in blasts of atomic bomb dimensions. Naturally, the project was shelved, although the advice was totally wrong. Paradoxically, this ridiculous state of affairs did not end there. Soon afterwards, another foreign 'expert' who was totally ignorant of the facts of volcanism persuaded the authorities that they could halt the lava that was threatening to block the entrance of the harbour and destroy their town, by sprinkling large volumes of water onto the hot surface from fire-boats and banks of land-based monitors. Once again, it was unsound advice that was adopted. As Haroun Tazieff recalls: 'No argument could prevent the exercise: not even the evidence that the Atlantic Ocean itself, with all its water, had been unable to stop the main part of the flows which had crawled over the sea floor for two months.' Experienced volcanologists are familiar with the relatively quiet, natural entry of great flows of incandescent lava into the oceans and know very well that even the weight of countless millions of tons of water, many thousands of feet deep, does not prevent lava flows extruding onto the ocean floors in sufficient volume to form the mid-ocean ridges and to build up volcanic islands like Surtsey, Deception Island and Tristan da Cunha. Ill-conceived, inexpert advice, which involves a small country in needless and sometimes astronomical expense, is both cruel and criminal.

Another event of this nature took place during the 1976 eruption of the Soufrière of Guadaloupe, West Indies, when skilled volcanologists who were investigating the eruption at the time reported that there was no real danger from the activity. It was their considered opinion that the greatly feared glowing avalanches of nuées ardentes which might have come from this volcano were very unlikely to occur as a result of the type of eruption in progress. This opinion was based upon many years practical experience of volcanic eruptions and their observation that no fresh magma was being released in the crater: in fact, the ash being thrown out by the eruption was no more than reworked older volcanic debris. Despite this report from a most experienced volcanologist, supported by careful studies of the volcano made by his research team, the views of two other 'experts', who were devoid of volcanological skill, were unwittingly taken into consideration by the civil authorities and, during a temporary absence of the volcanological team leader on another volcano, these 'experts' brought about a total evacuation of 73,600 people at enormous cost. One of these other scientists was later forced to reveal that he had produced false evidence in his report, incorrectly suggesting that fresh magma was being erupted. Intellectual dishonesty and propagation of unsupported conclusions had led to unnecessary panic evacuation and to a very skilled scientist being unfairly castigated and removed from office, albeit temporarily.

These cases illustrate the importance of choosing, well in advance, suitably qualified advisors to deal with and investigate earthshock problems of all kinds. The scientist who is most qualified on paper and shouts the loudest is not necessarily the best man for the job. What is needed is a specialist with first-hand experience of disaster team work and procedures, who is able to assess the situation calmly, who is responsible, skilled in his own particular field of knowledge and able to correlate and interpret *all* the relevant data and other expert opinion before giving any definite advice. His prime consideration must be the interests of the people for whose lives he may be indirectly responsible; in any disaster the restricting conventions of 'pure' scientific research must take second place.

Tremendous advances have been made in the study of earthquake risk and hazard in recent years. Earthquake risk maps, showing the likelihood that damaging earthquakes will occur at any locality, have been produced for many countries: instrumental techniques are available to measure and monitor any dangerous build up of strain in the rocks of the Earth's Crust within earthquake-prone zones; the special design and constructional requirements of buildings which must withstand earthquake shock are clearly understood and it is known which land must be avoided when building in these zones. Yet, throughout the world, people are still living with inadequate protection in areas of very high earthquake risk, unnecessarily exposed to well-understood hazards, and buildings are still being constructed on land known to be dangerous, because it is cheap. When one learns of hospitals, emergency relief centres, fire and police stations all having been sited on land exceptionally prone to earthquake hazard, it is hard to believe that mankind can be so collectively foolish.

In 1949, more than 10,000 people were killed when an earthquake in Tadzhikistan triggered a series of fatal landslides. The Soviet Union despatched a team of scientists to study this remote area: they were asked specifically to search for any symptoms that might warn of future earthshocks. After 20 years of painstaking work the Soviet team revealed their findings at an international conference. They had discovered that, in the period before an earthquake struck, there were measurable changes in the electrical conductivity of the rocks deep in the Crust; a greater concentration of the radioactive gas radon was found in water from deep wells; the land surface above a potential earthquake focus was distorted both vertically and horizontally and there was a distinct variation in the velocity of seismic waves traversing the stressed rocks, a variation that could be identified using conventional seismographs. Japanese, American and other scientists had been working towards similar conclusions and, by pooling world knowledge, the exciting possibility that earthquakes might be predictable became a feasibility for the first time in the early 1970s.

Today, we know that earthquakes can be predicted *if the proper information is available*. In earthquake-prone areas, such as those

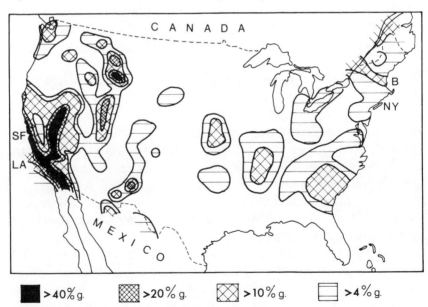

Fig. 35   Earthquake hazard map for the United States showing the amount of ground shaking, expressed as a percentage of gravitational acceleration in the horizontal, that could occur within the next 50 years. The maximum value along the San Andreas fault system is 80 per cent of g. SF – San Francisco, LA – Los Angeles, B – Boston, NY – New York.

adjacent to the San Andreas fault in California, the 'pulse of the Earth' can be continuously monitored in a number of ways. Surveyors, using conventional levels and theodolites, and a number of sophisticated tools, including tiltmeters, creepmeters and laser ranging instruments, can map the gradual distortion of the ground. Magnetometers, gravimeters, seismographs and resistivity gauges can be used to check slow changes in the characteristics of the rocks far underground and scintillation counters will signal changes in the amounts of radon gas dissolved in well waters. Many of the physical changes that are observed by these methods are due to dilatency – rock under extreme stress expanding with myriad cracks which allow water to enter. Chinese scientists have pioneered some additional methods of earthquake prediction based upon their studies of the unusual behaviour of various animals immediately before a large 'quake. Since 1970, as a result of all this work, there have been one or two notably successful earthquake predictions in America, the USSR, Japan and China, but against this very limited success must be listed the many unpredicted disastrous killer earthquakes that have occurred during this period throughout the world. Obviously, a great deal of further knowledge is required and considerably more effort and money must be put into the gathering of detailed information in earthquake zones before the tremendous potential and promise shown by the few successful predictions can be fully realized. If the scientists involved in this work are given sufficient encouragement and adequate financial support there is no reason why earthquake prediction should not become a routine matter within a decade. What is required is a worldwide earthquake prediction service that can provide the expertise necessary to save lives in both the developed and undeveloped countries of the world, for this expensive and technologically exacting procedure is at present beyond the capability of many of those countries in which earthquake risk is greatest. Slow progress is being made towards the provision of such a service, but the present international measures in this field are still very far from adequate.

Satisfactory protection from earthquake hazards, however, requires much more than accurate risk evaluation and successful earthquake predictions. Obviously, continuing research is necessary into what kinds of land to avoid, what kinds of buildings should be constructed and when and where earthquake shelters are necessary. Equally important is that the authorities should adopt a responsible attitude, employ the necessary specialists, take their advice more seriously and provide them with adequate funds. Proper information must be made available to the general public and people must be prevented from falling victim to their own or other people's stupidity and avarice by strictly enforced laws. Any kind of construction must be forbidden on land through which known faults pass, land prone to earthslide, land over water-sodden sediments or land known to have suffered in previous earthquakes. In coastal areas at risk from tsunamis, the provision of tsunami warning systems and tsunami shelters, together with adequate procedures for rapid evacuation, should be compulsory. Such a warning system is now operative in the Pacific, where, as a result of the incorporation of a network of Soviet warning sites with those already operated by Japan and the United States, a comprehensive 'multi-nation' tsunami alert scheme provides advance notification of the approach of giant killer waves. It is to be hoped that from this promising start a worldwide tsunami alert system may develop.

One of the most exciting research developments of recent years is the suggestion that it might be possible to prevent the recurrence of disastrous earthquakes along known faults by encouraging the accumulating rock strain to dissipate itself continuously rather than only occasionally and with great violence. In 1966, US Geological Survey scientists noticed that minor earthquakes were being triggered around Denver, Colorado, by the high pressure injection of industrial waste fluids into the sub-surface. After a series of experiments lasting for some years, they concluded that the carefully controlled injection of high pressure fluid into potentially dangerous fault zones could increase the dilatency of the rocks and 'lubricate' the fault planes to such an extent that violent movements separated by long intervals of time could be replaced by a continuous or near continuous creep over selected lengths of the fault, thus eliminating all earthquakes of sufficient intensity to interfere with human life. Hopefully, this attractive hypothesis may be put to a practical test in the near future.

While earthquakes kill effortlessly and destroy property without great displays of the natural forces involved, the same certainly cannot be said of volcanoes. Volcanic hazards may be divided into three major categories: lava flows, pyroclast falls (ash falls) and pyroclast flows (ash flows, nuées ardentes, glowing avalanches, base surges, volcanic mud flows and related phenomena). There can be no doubt whatever that a large volcano in full eruption is the greatest display of natural energy and the forces of nature of which the Earth is capable. When such forces are directed against densely populated areas, perhaps due to man's indiscriminate development and poor planning, the outcome can be utterly devastating. In spite of this, however, it is now possible for skilled volcanologists to recognize those volcanoes which can give rise to such horrors as pyroclast flows and to delineate zones of maximum possible danger. Thus, with proper use of our present knowledge of volcanology there is much that can be done to minimize the death toll and the suffering that volcanic earthshocks can cause.

The most serious lava flows are those which result from flood eruptions, where large volumes of fluid lava are poured rapidly out of a long fissure to cover vast areas; the 1783 activity of Laki in Iceland was such an eruption. Damage to agricultural land by lava flood is total. Other kinds of lava flow can be equally destructive: they form more or less discrete streams and can travel considerable distances, cut communications and destroy property. However, destruction by lava flows can be minimized in a variety of ways: the flow can be diverted into adjacent valleys where it will cause less damage; the lava stream can be bombed or breached with explosives to help change its direction; or, alternatively, channels can be excavated round villages to direct the paths of possible lava flows onto land of lesser importance. In Sicily, however, neither of these procedures is possible, for there is an ancient law which forbids the diversion of lava flows. The law dates from 1669 when, during the largest recorded eruption of Mount Etna, in which reputedly 100,000 men, women and children died, a great lava stream was advancing upon the city of Catania. A certain Diego Pappalardo organized an intrepid band of followers who, protecting themselves against the heat with wet cow hides, dug through the side of the lava flow with green poles. Their efforts were successful, the lava was diverted through the breach that they had created and the main flow that was threatening Catania slowed down, but their activities had created another problem, for the diverted lava flow then advanced towards the small village of Paterno and several hundred citizens of this village, roused to fury by the proceedings, attacked and drove off Pappalardo and his men. Unattended, the breach soon healed up and the main stream continued down to Catania and eventually destroyed the western part of the city. Since then, the diversion of Etnean lavas has been forbidden. The attempt that was made on Heimaey in 1973 to halt a lava flow by spraying it with water did cause a crust to form on the lava, but it certainly did not halt the flow: only cessation of the volcanic activity saved the partly buried town of Vestmannaeyjar from further destruction and possible total burial. The overriding criterion in these cases is that any preventive measure which involves diversion of lava flows requires a favourable topography for success. Research in volcanology has now progressed to the state where, within 48 hours of the birth of a new lava stream, we are able to predict the distance to which it is likely to flow, at what time it will reach given points and the course it will follow. During recent lava eruptions we have made such predictions with a 95 per cent success rate. Lava flows only rarely cause extensive loss of life, but the destruction to property is total. Agricultural land covered by lava takes a long time to recover, with the exception of that located in tropical humid climates, where the processes of weathering are accelerated.

More serious than lavas in their effect on life and property are the pyroclast falls which occur when gases carrying fragments of gas-rich magma that have expanded into a froth are expelled rapidly from the throat of a volcano. This froth solidifies to form pumice and ash. Prevailing winds may carry the pumice and ash enormous distances and deposit them far from the volcano along the dispersal axis. Light falls deposit a thin covering of ash over the surrounding countryside, chiefly damaging vegetation; heavier falls from violent strombolian activity can strip trees of their leaves and branches, and can have profound effects if near a town. When the eruption of Eldjfell volcano on Heimaey partly buried the nearby town in 1973 (Plate 12), the strombolian ash fall was sufficiently heavy to cause buildings with flat roofs to collapse under the weight of the ash. Others were structurally damaged and quite beyond repair. In this instance the people of Heimaey were fortunate, for they were able to excavate part of their buried town from beneath the ash cover after the eruption had ended.

The heaviest ash falls are produced by plinian type eruptions. These very destructive ash falls cause the roofs of most buildings to collapse, but often leave the walls standing. Plinian eruptions release enormous volumes of ash in a comparatively short time, and this hampers communications, hinders evacuation and impairs the function of both machines and men. All ash falls can damage the teeth of grazing animals and clog their digestive system, and the poisonous elements that volcanic ash sometimes contains can so contaminate the ground as to render the grass lethal to livestock. Ash falls may also upset the ecological balance so that one species may be killed to the benefit of another that is more hardy and agriculturally destructive. This happened during the eruption of Paricutin volcano in 1943, and resulted in the destruction of valuable sugar cane crops by cane boring beetles. Fires caused by lava flows are a serious hazard during eruptions, for they can destroy forest land and timber buildings; other fires can be started during ash fall activity by red-hot volcanic bombs crashing through roofs of houses and igniting flooring and furniture – a common occurrence during the eruption on Heimaey.

Recovery from the effects of ash falls is relatively quick, and under some circumstances it may be possible to excavate buried homes. The material presents a large surface area to weathering processes, and under the right climatic and agricultural conditions will quickly break down to yield extremely fertile soils; it is for this reason that many dangerous volcanoes are heavily populated on their lower slopes. People tend to move back to their old farms after an eruption in order to cultivate any land that is left and to plant as soon as possible on the new ash and lava fields. Perhaps the answer is to legislate to prevent housing resettlement in areas that are repeatedly devastated while at the same time providing good access to enable farming to be carried on from a distance once the land has recovered. At least this would minimize further loss of life and damage to buildings during subsequent eruptions. Of course some plinian eruptions are so big, and devastate such huge areas, that this would not be possible; nevertheless, there are many volcanoes throughout the world where such legislation would ultimately pay off in both human and economic terms.

Pyroclast flows – glowing avalanches or nuées ardentes – the most devastating of all types of volcanic activity, cause total destruction to life and property. Land, however, is easily reclaimed after these eruptions and, as in the case of heavy ash falls, people tend to move back into known danger zones, quite oblivious to the fact that it can all happen again. Death during glowing avalanches is particularly horrible; it results from instant cadaveric spasms due to inhaling red-hot rock dust; other victims may suffer asphyxiation, severe burns or burial by fallen buildings. Only well-constructed brick buildings are likely to survive, though this is not always the case; as we saw in Chapter 5, St Pierre town on the island of Martinique was completely levelled by a nuée blast in 1902. Widespread damage also occurs from secondary fires, and enormous casualties can result from associated secondary effects such as disease and famine. Glowing avalanches are particulate flows composed of a fluidized mass of red-hot rock particles. It is for this reason that they behave as liquids and flow easily over the countryside following river valleys. They are highly mobile, travelling at speeds of over 100 miles per hour on the steep slopes of the volcano and slowing down to perhaps 40 miles per hour over flatter ground on the lower slopes.

The nuée at St Pierre covered the 4.4 miles from the vent of Mont

Pelée to the town in three minutes during the eruption in 1902, an average speed of 87.5 miles per hour; thus no warning was possible once the nuée had been released. Despite the apparent finality of these dusts of destruction, there have been some incredible cases of survival. During the 1951 eruption of Mount Lamington in Papua/ New Guinea, nuées swept down river valleys killing almost all who stood in their path. However, where groups of ten or more people were huddled together in terror those on the outside often protected those in the centre of the group, some of whom survived. In one instance, a man who saw such a cloud bearing down on him jumped into the river and dived beneath the surface; when he could hold his breath no longer he surfaced, only to find the heat unbearable and, despite his burnt mouth, face and back, he dived once more beneath the water. The next time he surfaced the danger had passed! Those who have experienced a glowing avalanche and lived to tell the tale speak of a feeling of intense heat and suffocation during the few moments when the fire cloud is passing.

The zones affected by nuées are always dangerous for some considerable time after the first flow, because several subsequent nuées can be expected from the volcano. Later in these eruptions a lava dome or spine often develops in the vent. This should always be kept under careful surveillance, for the area is not safe until it can be established beyond all doubt that the dome has finally ceased to grow. It is very doubtful if man will ever be able to combat the very largest type of glowing avalanche eruption, the type that can produce immense thicknesses of ignimbrite (rock deposited from a pyroclast flow) extending many tens of miles from a volcano. Such flows are known to have overridden the mountains east of Naples, for example, and completely buried immense areas under many feet of hot ash. Southern Tenerife is also covered by many square miles of ignimbrite which was deposited by large eruptions from an old vent in Las Canadas. The largest ignimbrites known are hundreds of feet thick and may have covered areas over a hundred miles across. Many such enormous pyroclast flow eruptions have taken place in the past and there can be little doubt that similar devastating events will reoccur at some time in the future. Total evacuation of areas at risk is the only sensible procedure once it is clear that an eruption is imminent.

Mudflows (lahars) are yet another hazard that may be caused by heavy rainstorms during large eruptions. The loose hot ash which covers the slopes of a volcano is easily turned into a boiling mudflow under conditions of torrential rain (the same effect may be brought about by rain falling on nuées, or indeed by nuées entering large bodies of water, such as rivers). Mudflows travel in much the same way as glowing avalanches, but are more easily affected by the topography, which means that under some circumstances they could be diverted. Their speed, though slower than nuées, is none the less high enough to make them dangerous, and their high mechanical strength gives them great destructive power: boulders 20–30 feet across are carried easily by such flows. An eruption in a crater lake is always very dangerous. Mudflows can quickly form, and if the lake contains highly acidic water the ash falling from explosions within the lake can cause severe burns, as in the eruption of Kawah Idjen Volcano on Java in 1936.

Another hazard which sometimes accompanies large volcanic eruptions in shallow seas is the creation of giant tsunamis, which can wipe out entire coastal areas with tremendous loss of life. As we have seen, the great death toll of Krakatoa in 1883 was mainly due to this cause.

Emission of poisonous and asphyxiating gases such as carbon dioxide can sometimes accompany volcanism. These accumulate in hollows and in the cellars of buildings and have claimed many victims in the past. In 1947, carbon dioxide accounted for the death of numerous sheep during the eruption of Hekla Volcano, Iceland, and one man is known to have died as a result of noxious gas during the eruption on Heimaey in 1973.

In spite of the enormity of these hazards there is much that man can now do to prevent the great loss of life and to alleviate the suffering that volcanic eruptions cause. The most useful and rewarding task for the volcanologist is to study the deposits of past eruptions, for here is the record of a volcano's activity and also the best guide to how it is likely to behave in the future. Once the types of activity that characterize a particular volcano have been recognized, zones of maximum danger can be delineated. When an eruption is due – prediction of volcanic eruptions is dealt with later – contingency plans can be put into operation. If a particularly big eruption is expected then the entire area should be evacuated. For smaller eruptions only the known danger zones need be cleared. Should glowing avalanches be expected, then alterations to existing topography, although never tried, may help divert flows. Plans for such modifications should, however, be put in hand well before an eruption is due to occur. Villages which are on the banks of, or in direct line with, a valley on a volcano are especially vulnerable. If an adjacent valley exists then it may be possible to excavate and dam the first valley so as to divert the volcanic stream from this valley to the next. This could effectively

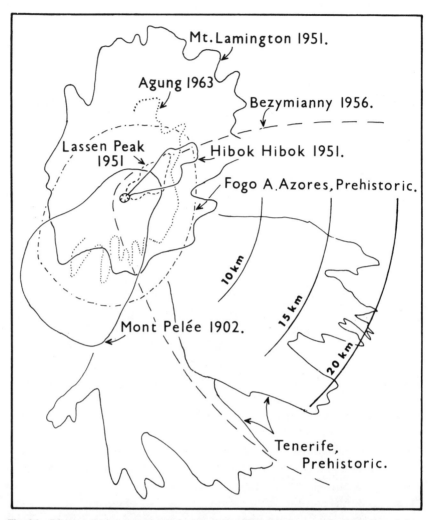

Fig. 36 Diagrammatic comparison of some areas affected by large-scale volcanic eruptions. Three arcs showing distances of 10, 15 and 20 km from the vent allow the aerial extent of the damage zones to be assessed. Some Andean pyroclast flows and prehistoric Caribbean and Central American pyroclast falls have greatly exceeded even this scale.

channel small nuées, mudflows and lava flows into areas of lesser importance. Villages in known danger zones should be provided with adequate underground shelters which will protect the population while a nuée is passing, and careful thought must be exercised during building construction. Roofs, for example, should be steeply sloping (at least 35°) to prevent the accumulation of ash, and they should preferably be made of strong non-combustible material to prevent hot fragments penetrating and setting fire to the house interior; windows should also be protected for the same reason. Livestock is particularly vulnerable during volcanic eruptions: facilities should exist for their evacuation to places of safety in times of danger. Evacuation plans are also needed for coastal areas that might be swept by tsunamis. For notoriously dangerous coasts, legislation could prevent building there. Crater lakes can sometimes be drained and the water kept at a safe level – this has been done at Kelut in Java – despite the fact that the drainage system is destroyed during each eruption. This simple procedure could prevent great loss of life from mudflows.

Perhaps the best way to combat volcanic hazards is to prepare both general and specific predictions of activity, and to take protective measures as we have outlined above. An experienced volcanologist should always be on the spot to advise as an eruption progresses. During the last decade, we have learned a great deal about how volcanoes behave and what type of eruption to expect next in a volcanic area; this knowledge is now being consolidated by active research. The time has certainly come when we should be applying our knowledge of volcanoes for the benefit of mankind. Because of the natural association of man with the highly fertile agricultural areas in which volcanoes are plentiful, there is a real need for at least a minimum degree of surveillance. Those volcanoes which are near densely populated areas deserve rather special attention, as any particularly large outburst could be disastrous and have far reaching social and economic consequences. Preparation and publication of volcanic risk and hazard maps would at least indicate the extent and area of presumed damage, but accurate surveys are required to indicate what kind of eruption to expect next in a volcanic area and to allow authorities to legislate to control building in such areas. Unfortunately, it is these dangerous areas which attract people, as the soil is rich; volcanic ashes and rocks decompose to provide some of the most fertile soils of the world and people who depend on agriculture for their livelihood naturally are attracted to the fertile slopes of large volcanoes. In spite of the destruction caused by each eruption on the Vesuvius and Etna volcanoes in Italy, people move back onto the slopes after each eruption to carry on their time-honoured lifestyle. During long repose periods farms are built and plots are cultivated progressively higher and higher up the slopes. Eruptions similar to the AD 79 event on Vesuvius or the AD 1669 event on Etna could cause an enormous loss of life in the future. It is hoped that, today, most civil authorities are sufficiently aware of the problems to allow speedy evacuation to take place once a volcano has embarked on a sequence of events that will culminate in disastrous eruption. Only painstaking attention to detail during a volcanic survey of a suspect volcano will reveal the kind of eruptions the volcano is capable of giving. Such important work really comes under the heading of general prediction: it involves full geological interpretation of all the deposits from ancient eruptions, for each ash layer is like the page of a book telling the history of the volcano, and the field investigations must be coupled with analysis of the records of all past activity. Specific prediction may be more precise, but is very costly and utilizes expensive instrumentation and involves the analysis of geophysical data from suspect volcanoes. A general volcanic survey will give a rough guide to the future behaviour of a volcano based on the premise that the past behaviour is the best guide to how it is likely to behave in the future. Generally, a volcano with a long history of violently explosive eruptions, separated by long repose periods during which, perhaps, there were many similar smaller events, is a suspect volcano. If it has a long and violent past then there is a high probability that it will continue to behave violently in the future. The most useful feature about general prediction is that it alerts volcanologists to the dangers from possible eruptions and allows time for the installation of geophysical instruments to monitor the volcano more closely using specific prediction techniques.

For some volcanoes, the record of past activity is sufficiently complete to permit the volcanologist to calculate a crude periodicity. On a short-term basis such analysis has proved very accurate in predicting the times of explosions during eruptions. But over longer periods the predictions become vague unless augmented by other, more specific, techniques. Unfortunately, only a small number of volcanoes in the world are sufficiently well documented to permit any kind of general prediction to be made. Apart from those for which the historical record is good, the only way to ascertain if the volcano has a long history of violent activity is to conduct a thorough volcanological study of the area. Every deposit from past eruptions must be carefully logged and correlated with other deposits elsewhere on the volcano, and only then is it possible to draw up hazard/risk maps, to calculate the volume and hence the magnitude of past eruptions, and to arrive at a rough periodicity for the volcano's activity.

Such a thorough study will also reveal considerable details about past eruptions; whether it was raining at the time, whether mudflows or pyroclast flows descended the volcano slopes, whether the eruptions were accompanied by violent lightning displays and whether the slopes of the volcano collapsed inwards during quieter periods, only to be blown out again when activity recommenced. Careful mapping

of ancient volcanic deposits can give valuable information about wind
direction and speed, the velocity at which the material was ejected
(the muzzle velocity) and the height of the eruption cloud. It can tell
us whether the blasts were angled or vertical and the probable depth
and pressure at which the explosion took place. Such investigations
have shown, for example, that the Tenerife volcano in the Canary
Islands has a long history of extremely violent eruptions. During the
past 30,000 years there have been over 15 pumice-fall eruptions
around the volcano, some of them of considerable magnitude and
accompanied by at least an equal number of large pyroclast flows
formed from incandescent volcanic sprays. If we go even further back
in the history of that volcano, an additional 30,000 years, we find that
over 21 pumice-falls and more than six pyroclast flows occurred; this
does not include the many smaller ash and mudflows that also took
place. Some of the prehistoric eruptions on Tenerife were as large in
magnitude as the celebrated 1883 eruption of Krakatoa. At present
the volcano is live, but dormant, with active fumaroles in its summit
crater.

In the Azores, the island of San Miguel has a long history of violent
activity, with as many as 26 eruptions during the past 5000 years, 10
of them large. Guadalajara City in western Mexico is built on the
deposits from the nearby La Primavera volcano, which has given as

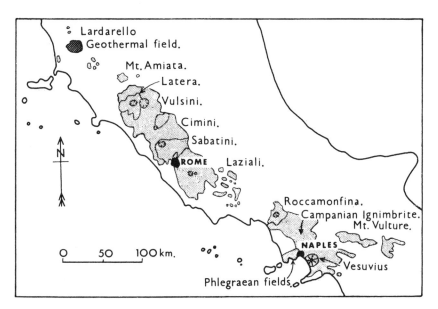

Fig. 37 Central Italian volcanoes showing the areas affected by past explosive eruptions from
volcanic centres with exceptionally long repose periods. Not only is Rome considered to be in
danger from a future event, but Naples also.

many as 12 particularly violent pyroclast flow and pumice-fall eruptions, some in an ancient crater-lake. Similarly, in central Italy, there
are many thick ignimbrites, the result of pyroclast flows from prehistoric eruptions of enormous magnitude. Rome, for example, is built
on one such ignimbrite deposit. None of the foregoing volcanoes can
be considered extinct by any criterion: they simply have very long
repose periods, sometimes as long as 5000 and perhaps exceeding
10,000 years.

In the case of specific prediction, the geophysical techniques used
rely on the fact that, prior to an eruption, large volumes of magma are
introduced into the base or superstructure of the volcano. This
magma disrupts solid rock as it is introduced and the heat which the
magma releases causes changes in the physical properties of the
surrounding rocks. Volcanologists can measure these changes and
detect the earthquakes which the ascending magma causes. Measurements must be carried out over reasonable periods of time, so that
geophysical observations can be matched to visual ones. One technique is to use sensitive seismographs to measure the tiny earthquakes
that occur when a body of magma prises solid rocks apart as it forces
its way into the base of a volcano. Nearly all volcanic eruptions are
accompanied by earthquakes, although the converse is not true. With
careful measurement it is possible to know not only the course the
magma is taking on its way to the surface, but also the rate of ascent.
By this means it is possible to tell where and when an eruption is due.
Not all ascending magmas reach the surface to cause an eruption.
Sometimes the magma stops rising, solidifies, and forms an intrusion
instead.

As magma forces its way up into the base of a volcano, it causes the
structure of the volcano to swell. This tumescence or volume increase
which can be measured accurately with tiltmeters is often accompanied by seismic shock swarms. On some volcanoes such swarms can
increase to over 1000 per day. During the 1955 eruption of Kilauea
volcano on Hawaii, seismic shocks averaged 15 per day between 1
and 23 February; by the 24th the number had increased to 100 per
day; two days later it reached 600 and the following day 700. On 28
February the frequency of shakes increased still further and at 8.00
on that day the eruption started. The forecast that an eruption was
due to start was not based solely on the data from one type of
instrument, however. Rarely is that sufficiently accurate. Data from
tiltmeters which were accurately measuring the tumescence of the
volcano augmented seismographic records.

Tiltmeters are sensitive spirit levels which can measure a movement of as little as 1 mm in 1 km. Basically, they consist of two fluid reservoirs connected by a long tube. Any movement of one reservoir relative to the other is indicated by a change in the level of the liquid in both reservoirs which can be measured accurately. Two tiltmeters placed at right angles are all that is required to define the direction and amount of tilt. During 1958 and 1959, tiltmeters on Kilauea volcano indicated a steady swelling of the summit region. By November 1959 the rate of tumescence had more than doubled and by 14 November volcanic earthquakes in the summit region had increased in intensity and number. At 8.00 p.m. that day, lava gushed from a crack in the wall of Kilauea Iki crater. With the loss of this lava the tiltmeters recorded a sudden detumescence in the summit region, but within a week renewed tumescence indicated further activity. The eruption started again on 26 November from the summit vent and continued until 21 December. At this stage tiltmeters indicated not a subsidence but a tumescence even greater than when the eruption first began. Volcanologists forecast another eruption at a point 25 miles east of the summit. On 13 January 1960 an eruption took place from this point and 150 million cubic yards of lava poured down the hillside, over a cliff, into the sea. This activity was followed by a rapid detumescence of the summit area.

Distance measurements can augment tiltmeter information. As a volcano tumesces, the distance between fixed points increases and these distances can be measured accurately with special instruments using laser beams. The altitude of bench marks is also changed by tilt and can likewise be monitored. When magma is injected into the base of a volcano it may heat up the surrounding rocks above their so-called Curie temperature and thus demagnetize them. This causes a change in the magnetic field of some volcanoes which can be detected with a magnetometer. During eruptive cycles on O-Shima volcano, Japan, the local magnetic dip can increase by as much as 30 minutes of an arc. The injected magma may similarly cause a change in the local gravitational field of a volcano and regular surveillance with gravimeters may reveal such a change – although it is normally very small compared with the accuracy of existing instruments. Chemical monitoring of the composition of volcanic gases (Plate 26) on certain volcanoes is a technique of limited value at present, and as such can only supplement existing geophysical techniques. It does, however, hold great promise for the future.

Man is dependent upon volcanoes, more so in some parts of the world than others. Volcanoes provide fertile land, mineral resources such as porphyry copper deposits, lightweight building aggregate and sometimes geothermal power. To live in the presence of a potentially dangerous volcano has, for many people, become commonplace. There are many cities around the world which exist within 'reach' of volcanoes which are live but dormant: the city of Portland, Oregon, lies at the foot of Mount Hood; Guadalajara City, a few kilometres from La Primavera; Arequipa City, Peru, at the foot of El Misti volcano and, as we have seen, Rome is within reach of potentially dangerous volcanoes and during an eruption could vanish overnight. The risk, admittedly, is hypothetical, but it is a geological certainty that volcanoes such as these will erupt at some time in the future. It is doubtful if the tremendous death toll which will undoubtedly ensue will be averted by prediction and evacuation. Too few countries are sufficiently alert to the dangers, and the official attitude is generally 'It can't happen here'. But it can. Perhaps only a volcanic earthshock will convince governments of the value of categorizing volcanoes and establishing monitoring schemes.

If, after a consideration of earthquake and volcanic hazards, there is a strong case for setting up an international body to oversee and co-ordinate our disparate national effort at mitigating the effects of natural disaster, then it becomes overwhelming when the possibilities of future earthshock from extraterrestrial bombardment are considered. Let us, for the moment, consider a possible sequence of events that might occur should a really large extraterrestrial object approach the Earth today. It is very likely that the first sighting of such a 'rogue asteroid' moving towards us on a possible collision course might be made by one of the many enthusiastic amateur astronomers or 'star gazers' who make watching the various members of the Solar System their hobby. Initially, lurid publicity regarding the discovery and its possible dire consequences, as revealed in uninformed discussion in the press and on television, would inevitably result in a bout of fruitless scientific controversy and denial, with some experts dismissing the observations of a mere amateur whereas others, while confirming his observations, would come to very different conclusions regarding the consequences. A few religious sects might immediately foresee the total destruction of the world as foretold in their own predictions and dogma. Meanwhile, the official space agency astronomers – after an inevitable delay before it was decided that this matter was of sufficient importance for them to be redirected from their routine studies of each other's artificial satellites and space probes and their regular watch for the launching of 'enemy' missiles, spy and killer satellites – would have been reporting in secret to their respective governments. Once the reality of the situation had been accepted by their respective senior government scientists and they had persuaded their political overlords that a genuine danger existed, what might the next move be? The events of 1978, when a Soviet nuclear satellite crashed onto Canada, suggest that a cloak of secrecy would be their first reaction.

The news would obviously be political dynamite on a world scale. It would be clear immediately that if one of the superpowers were to be

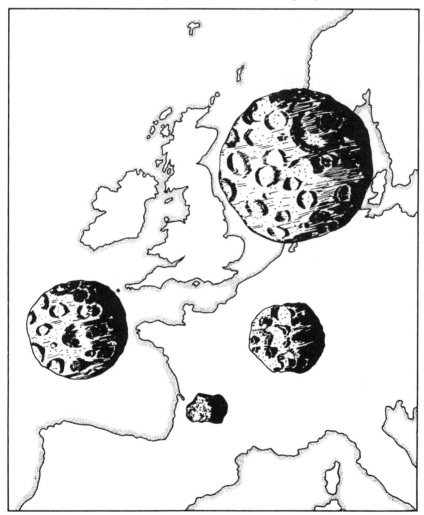

Fig. 38   Four of the largest known asteroids: Ceres (480 miles in diameter), Pallas (300 miles in diameter), Vesta (240 miles in diameter) and Juno (120 miles in diameter). A collision with any of these monstrous chunks of solar debris would completely destroy and devastate Europe. Collision with quite small asteroids would be cataclysmic in a populated area. For comparison, the Earth is 7920 miles in diameter and the Moon has a diameter of 2160 miles. In the figure, the four asteroids are shown to scale over the map of north-west Europe.

the recipient of a direct hit, the balance of world power would be changed dramatically. Politicians being what they are, their first reaction would be related to this concern, and emergency discussions with their close allies would have priority over public announcements or immediate East/West meetings to co-ordinate a world disaster response. Nevertheless, it would eventually become clear that an international response on a world scale was required. All this would take up valuable time, because there would be no precedents and no prepared channels for the dissemination of information and co-ordination of possible action. While the lengthy, behind-the-scenes political manoeuvring and final reluctant agreement to the matter becoming an overriding world concern were taking place, the Earth and the rogue asteroid could be approaching the impact position at a high speed. The world's press and television would suddenly become aware of the true gravity of the situation; even the most casual members of the public would begin to realize the incredible danger that faced mankind and, in many places, widespread panic and resultant public disorder, looting, rape and drunkenness would inevitably ensue. During the last few hours or days before the impact the hastily convened World Disaster Committee, consisting of scientists, military men and politicians, would be in permanent session, weighing the futility or otherwise of firing every nuclear device available at the asteroid against the possible massive and even fatal contamination of the environment that this might cause after the collision, even if it did result in the break up of the extraterrestrial object. Some would regard the danger from the multiple impact of numerous highly radioactive fragments as greater than that of a single large impact without additional nuclear contamination. The probability is that, in the absence of previously planned and carefully thought-out responses, the committee would take ineffective action too late and the rogue asteroid, partially fragmented and contaminated by numerous poorly co-ordinated hits by nuclear missiles would continue on its remorseless trajectory and strike the Earth with tremendous violence.

Of course, with some missiles from space this final eventuality may be inevitable, and evacuation of the impact danger zone the only possible action that mankind can take. Nevertheless, given sufficient time and a previously prepared plan, the possibility does exist that a small asteroid could be diverted from a collision course into one which would capture it as a new satellite of Earth, with all the attendant benefits this could bring – a nearby space platform for rocket launching, remote sensing, astronomical observations and solar power stations as well as a possible source of valuable metallic ores.

Man controls the future to a larger extent than perhaps many of us imagine. Setting aside such catastrophes as nuclear and biological warfare, man's deliberate, or for that matter, ignorant tampering with the environment could seriously affect his future and trigger a catastrophic natural disaster. Already scientists are becoming aware of the problems caused by the excessive burning of fossil fuels and escalating deforestation which together are responsible for the steadily increasing carbon dioxide level in our atmosphere, creating the so-called 'greenhouse effect'. If the global consumption of coal, gas and oil continues at its presently increasing rate then it has been calculated that the carbon dioxide content of the air could double over the next 50 years, with disastrous climatic side effects.

In Chapter 6 we discussed future frozen death or flooding in some detail. At certain times between 18,000 and 6000 years ago, world sea level was rising as fast as 20 feet per century. The expense of coastal defence against such a rapid rate of change in sea level, when it recurs some time in the future, as it undoubtedly will, would be

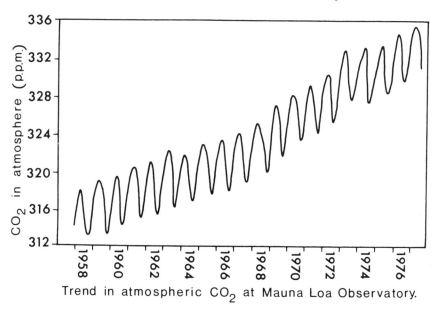

Trend in atmospheric CO_2 at Mauna Loa Observatory.

Fig. 39 Measurements made at the isolated, windswept Mauna Loa Observatory on the island of Hawaii show that the average carbon dioxide content of the atmosphere has risen by more than 5 per cent since 1958. The removal of carbon dioxide by photosynthesis during the growing season and its subsequent release results in a seasonal oscillation – dominated by the larger areas of plant-bearing land in the Northern Hemisphere – but it is clear that the average level of CO_2 in the air is rising at the alarming rate of about one part per million per year.

astronomical, and probably ineffective in the long run. The likely collapse of barriers more than 100 or 150 feet in height would eventually present an unacceptable risk to those living on land immediately behind them. For the past 6000 years, however, world sea level changes have been restricted to less than 20- 30 feet in total. Thus, as long as we continue in the current interglacial standstill period, and present indications are that this may last for anything between 100 and 10,000 years, effective preservation of the present coastline remains both worthwhile and feasible. In England, for example, the authorities are well aware of the problems which London faces in the near future. Subsidence of the Thames estuary relative to the level of the North Sea averages around 4–12 inches per century, and the configuration of the land areas bordering the southern part of the North Sea makes the eastern coastal area of England – Norfolk, Suffolk, Essex, London and Kent – particularly vulnerable to a combination of high tides and north-easterly gales. A barrier is even now being built across the lower Thames at Woolwich to control flooding by such future North Sea tidal surges. The present indications are that for some years it will do just that. But what would happen if unexpectedly heavy snowfalls in the Thames catchment area, with drifting up to 30 feet, as occurred in south-west England in February 1978, were to be followed by rapid thawing during torrential rainstorms and that these events took place a day or two prior to the barrier having to be closed to keep out a large North Sea surge driven by galeforce easterly winds? Clearly, any barrier across the Thames designed to keep the sea out would also act to keep the river in – so under these conditions London could rapidly flood behind its barrier! Calculations doubtless suggest that a coincidence of these factors is extremely remote, but so was a recurrence of the infamous 1953 east coast floods said to be at the time. Then, calculations were made which indicated that the same water level would be reached only once in every 200 years; yet the 1953 water levels were repeated in 1969, 1976 and 1978. The best available estimates have been shown to be wrong in the past; they could be wrong again.

The most satisfactory blueprint for the survival of London appears to lie not in small-scale, penny-pinching measures, but in the development of a system of polders like those that have been successful in the Netherlands. Putting aside cost for the purpose of the exercise, it is possible to envisage the erection of a system of bastions between east Kent and eastern Essex that would both protect the capital city of Great Britain for centuries to come, convert it into Europe's largest and most modern shipping port and establish the long-awaited third London airport away from populated areas; other bonuses would be the diversion of the main motorway arteries from the industrial Midlands and north of England to Europe away from London, an increased acreage of good agricultural land, constructive use of Thames silt, more efficient sewage disposal from the city and increased electricity production without the use of fossil fuels. The mouth of the Thames could be dammed by a construction, certainly no longer than that which stretches across the mouth of the Zuider Zee in the Netherlands, enclosing a new farming county and capable of enlargement should the necessity arise. A system of adjacent polders within this barrier, large enough to hold much more than the calculated maximum river discharge during high tide conditions, could be used for recreational purposes and partly emptied through a system of electricity generators at each low tide. These large polders could be connected to a series of smaller polders along the east coast of Essex and Suffolk into which sediment-laden water could be diverted during times of flood, to provide valuable new deposits of alluvium. Such ancillary polders could be periodically used as large settling tanks to contain treated sewage, and thus enhance the soil fertility; each would contain many smaller enclosures which could be flooded, drained and aerated for a year or so, and then flooded again on a regular cycle of between five and ten years. All main outlets could again use the discharge water to drive generators. Access to a new dock complex would probably be via a series of giant locks through a separate polder. The airfield approach could be over the sea, avoiding environmental disturbances as far as possible. The engineering problems of constructing such barriers are well known: certainly they are no more than those envisaged for the Bristol Channel scheme which was investigated recently by the Central Electricity Board.

Additional power could be obtained through the use of an external barrier of modified rocking float generators for extracting energy from the waves of the North Sea. Laboratory tests have shown that up to 95 per cent of the wave power can be extracted from artificially generated waves. Thus, an additional advantage of such a floating electricity power station would be the protective influence it would have on the coastal barrier if fixed some distance offshore, in maintaining an area of calm water between the floats and the barrier and thereby practically eliminating the effects of wave erosion and hence reducing maintenance costs on the dykes and polders. The ever-increasing costs of fossil fuels could soon make this proposition an economic reality, especially if the North Sea oil revenue were wisely used to fund such a scheme, and not frittered away. It would certainly help the unemployment problem and establish many valuable recreational centres.

In the early days of man's sojourn on Earth, individuals, family groups and whole tribes were killed by earthshocks of various kinds – storms, hurricanes, floods, fires, earthquakes, tsunamis, volcanic eruptions – but, because they were few in number and widespread over the face of the globe, the losses were small and the effects insignificant on the species as a whole. Only when large concentrations of people gathered in cities and intensive settled farming communities was it possible for great numbers to be killed by a single earthshock disaster. As long as the total world population remained small, and man retained his mobility of life as a nomadic hunter-gatherer and occasional farmer, he could survive through ice ages, desertification, changes of sea level and other similar disasters without serious loss, by simply retreating before the forces of nature, then readvancing as they became spent. The situation today is much less favourable. With the world's population at its present and projected levels for the near future, there is little room for manoeuvre. Retreat of large populations of the formerly ice-covered areas of the Northern Hemisphere into the tropics would be virtually impossible. Unless the world population trend can be reversed, the future earthshock prospect for mankind is horrifying: the larger and more densely concentrated the world's population becomes, the higher will be the inevitable death-toll in future earthshocks. Many planners foresee the spread of urban life over the entire land surface as a 'World City' – a smooth progression from metropolis to megopolis to ecumenopolis. A more ghastly and inhuman future could not be devised for, if the whole land surface of the Earth were to be as densely populated as London, New York or Tokyo are today, loss of life from unstoppable and uncontrollable natural disasters would run to several millions each year; a really major natural disaster would kill hundreds of millions. Surely, the callous acceptance of accidental death on such a huge annual scale – an inevitable necessity on such an overpopulated Earth – would be impossible, unless man were to become degenerated into a brutalized, mechanical subspecies?

Clearly, the answer to the question 'Can man and his planet, Earth, survive?' depends upon the time scale against which we examine the problem. In 5000 million years time, it is probable that the Sun will be dead and the Earth will have disappeared – but this prediction is of no direct interest to mankind today. Evolutionary studies tell us that our species, like every other species that has lived on Earth, must, after its allotted span, decline and be replaced by other, more successful life forms. Is man so unique a creature that he will prove to be exempt from this natural process? The answer must be no. Man as a species, and, indeed, *Homo* as a genus, will eventually become extinct.

The relevant question concerning man is not 'Will he survive?', but 'Could catastrophe severely curtail man's allotted span on Earth?' The answer to this question must of necessity be based upon many imponderables: it is, perhaps, partly dependent upon man himself and how he evolves socially over the next century. In view of the facts presented in this book, and available elsewhere, it seems that man must settle his differences, come to terms with the forces which surround him and genuinely collaborate on a global scale to predict where and when disaster will next strike. In addition, mankind as a whole must, at the same time, be ready to alleviate suffering in a region of major catastrophe by directing resources where they will do the most good. In some fields we have advanced a little towards this, but there is still a long way to go; and until such time as international co-operation is really forthcoming, there seems little chance of avoiding repetitions of the monumental mistakes, indecision and callous indifference of the past during major catastrophes of the future. Only by international co-operation can man curtail his own interference with the balance of nature and his misuse of the environment, conserve dwindling resources and exploit new and non-expendable sources of power. The state of the science of seismology has advanced now to the position where we are able to predict earthquakes in some areas and minimize their effects in others. Already volcanic eruptions can be predicted with a great degree of accuracy and all potential volcanoes can be zoned to assist authorities in legislating against building within the extreme danger areas. Many countries are still not equipped against abnormal flood conditions which cause so much death and deprivation. Is this the fault of the government or the Earth scientist? On the one hand government will, in its wisdom, decide that the financial burden of scientific investigation is too great, quite regardless of the valuable by-products which such investigations can give. The scientist, on the other hand, may be content to view events with detached scientific interest from the comparative safety of his ivory tower, where the risk of his academic reputation is not so great. Prediction and alleviation of global catastrophe demand involvement in the front line of the battlefield, not direction from the rear!

Only when we examine the possibilities of an ice age in the Northern Hemisphere, the wholesale flooding of the lowlands of the world or the effects of a large body from outer space striking the Earth, do we appreciate man's apparent insignificance. But how insignificant would man really be in the face of such impending disaster? Close co-operation between world scientists and the acceptance of their decisions by world governments and military authorities could, with careful evaluation, go a long way towards averting, or at least minimizing, a major world catastrophe. Pollution, atmospheric or otherwise, by man's thoughtless activities could, if unchecked, run riot and render the oceans and atmosphere of Earth extremely toxic in the cumulative sense. Already many disorders – brain damage, for example – have been proved to be the direct consequence of excessive pollution in our cities by the exhaust gases of the internal combustion engine. Legislation is under way in many areas to control this, but unless legislation against pollution of *all* kinds is enforced on a global scale the residual killer chemicals in the atmosphere we breathe, the food we eat and the water we drink, can only steadily increase. Perhaps the science fiction story of a select group of beings evacuating a heavily polluted planet to seek another sanctuary to pollute will prove to contain a grain of truth. Who, for example, 100 years ago would have dreamed that man would walk upon the face of the Moon, or for that matter in space itself?

Man as a species may have no more than a few million years before inevitable evolutionary extinction but, nevertheless, a proper knowledge of the dangers and hazards of life on Earth, coupled with a determination to work together in order to mitigate and avoid the worst consequence of both earthshock and our own folly, particularly runaway population growth, could perhaps prevent an unnecessary curtailment of our allotted span.

Index

14.95